Methods in Molecular Biology

Series Editor
John M. Walker
School of Life and Medical Sciences
University of Hertfordshire
Hatfield, Hertfordshire, UK

For further volumes:
http://www.springer.com/series/7651

For over 35 years, biological scientists have come to rely on the research protocols and methodologies in the critically acclaimed *Methods in Molecular Biology* series. The series was the first to introduce the step-by-step protocols approach that has become the standard in all biomedical protocol publishing. Each protocol is provided in readily-reproducible step-by-step fashion, opening with an introductory overview, a list of the materials and reagents needed to complete the experiment, and followed by a detailed procedure that is supported with a helpful notes section offering tips and tricks of the trade as well as troubleshooting advice. These hallmark features were introduced by series editor Dr. John Walker and constitute the key ingredient in each and every volume of the *Methods in Molecular Biology* series. Tested and trusted, comprehensive and reliable, all protocols from the series are indexed in PubMed.

Retinal Development

Methods and Protocols

Edited by

Chai-An Mao

Ruiz Department of Ophthalmology and Visual Science, McGovern Medical School at The University of Texas Health Science Center at Houston (UTHealth), Houston, TX, USA

Humana Press

Editor
Chai-An Mao
Ruiz Department of Ophthalmology
and Visual Science
McGovern Medical School at The University
of Texas Health Science Center at Houston (UTHealth)
Houston, TX, USA

ISSN 1064-3745 ISSN 1940-6029 (electronic)
Methods in Molecular Biology
ISBN 978-1-0716-0174-7 ISBN 978-1-0716-0175-4 (eBook)
https://doi.org/10.1007/978-1-0716-0175-4

This Humana imprint is published by the registered company Springer Science+Business Media, LLC, part of Springer Nature.
The registered company address is: 233 Spring Street, New York, NY 10013, U.S.A.

Preface

The vertebrate retina is an extension of the brain in the central nervous system. It has long been used as a model system to study the cellular and molecular mechanisms underlying the specification and differentiation of neural cell types, developmental construction and functions of neural circuitries, axonal outgrowth and pathfinding, and neurodegeneration and regeneration. Its external location allows for easy manipulation in vivo as well as isolation and growth in vitro. Additionally, it has evolutionarily conserved neuronal types and organization in specific nuclear layers; hence, mechanistic understanding from studies on one organism can often be applied to other model organisms. In recent years, it has become increasingly clear that in order to gain a comprehensive understanding of the design principles of retinal development, the establishment and function of neural circuitry in the retina, and the critical components for regeneration of damaged retina, it is necessary to combine multiple approaches that are traditionally used across different disciplines to understand the cellular and molecular mechanisms controlling each step of retinal development. The aim of this book is to provide interested readers with a set of practical experimental tools to study retinal development and regeneration and function of mature retinal neurons. Many of the protocols and strategies described in one organism can be easily adapted to applications in different model systems.

This protocol book includes several commonly used molecular and cellular techniques in molecular and developmental biology laboratories, as well as specialized methodologies for studying retina neuronal subtypes and electrophysiology. A number of model organisms, including chick, pigeon, South African clawed frog, zebrafish, and mouse, are included for different experimental designs. The designs and preparations of several commonly used molecular tools, such as adeno-associated viruses, transposons, and BAC transgenic constructs, are described in several chapters. Classical developmental techniques adapted for retinal tissues, such as the in vivo retinal regeneration system in *Xenopus laevis*, ex vivo mouse retinal slice culturing for imaging, three-dimensional embryonic mouse eyecup culturing for enhancing cell differentiation studies, and axon outgrowth assays, are included. For gene expression studies, single-cell RNA-seq strategies, sensitive RNAscope in situ hybridization techniques, and strategies for identification of *cis*-regulatory elements are described. Two chapters describe techniques for anatomical studies of retinal ganglion cell morphology and gap junction-mediated neuronal connection. The last two chapters are devoted to describing two specific electrophysiological techniques, multielectrode array recording on mouse retinas and paired recording to study the electrical coupling between photoreceptors.

I would like to express my deep appreciation to all of the colleagues who contributed their knowledge in retinal development and physiology and time to this book and to the staff at Springer for their patience and help. I also want to express my gratitude to Drs. John Walker and Kimberly Mankiewicz for their help in editing this book.

Houston, TX, USA *Chai-An Mao*

Contents

Contributors

CAVIT AGCA • *Lab for Retinal Cell Biology, Department of Ophthalmology, University of Zurich, Zurich, Switzerland; Molecular Biology, Genetics and Bioengineering Program, Sabanci University, Istanbul, Turkey; Nanotechnology Research and Application Center (SUNUM), Sabanci University, Istanbul, Turkey*

ANTHONY P. BARRASSO • *Department of Molecular Physiology and Biophysics, Baylor College of Medicine, Houston, TX, USA*

PUNITA BHANSALI • *Department of Ophthalmology and Visual Sciences, Albert Einstein College of Medicine, Bronx, NY, USA; Department of Genetics, Albert Einstein College of Medicine, Bronx, NY, USA; Department of Biological Sciences and Geology, Queensborough Community College, Bayside, NY, USA*

LAURENT BRODIER • *Department of Molecular Biology, Sciences III, University of Geneva, Geneva, Switzerland; Department of Biochemistry, Sciences II, University of Geneva, Geneva, Switzerland*

RUI CHEN • *Human Genome Sequencing Center, Baylor College of Medicine, Houston, TX, USA; Department of Molecular and Human Genetics, Baylor College of Medicine, Houston, TX, USA*

CHUAN-CHIN CHIAO • *Institute of Systems Neuroscience, National Tsing Hua University, Hsinchu, Taiwan*

SHUYUN DENG • *Advanced Technology Cores, Baylor College of Medicine, Houston, TX, USA*

RACHAYATA DHARMAT • *Human Genome Sequencing Center, Baylor College of Medicine, Houston, TX, USA; Department of Molecular and Human Genetics, Baylor College of Medicine, Houston, TX, USA*

RAVEN DIACOU • *Department of Ophthalmology and Visual Sciences, Albert Einstein College of Medicine, Bronx, NY, USA; Department of Genetics, Albert Einstein College of Medicine, Bronx, NY, USA*

WEI FANG • *Department of Ophthalmology, University of Pittsburgh School of Medicine, Pittsburgh, PA, USA*

CHRISTIAN GRIMM • *Lab for Retinal Cell Biology, Department of Ophthalmology, University of Zurich, Zurich, Switzerland; Zurich Center for Integrative Human Physiology (ZIHP), University of Zurich, Zurich, Switzerland; Neuroscience Center (ZNZ), University of Zurich, Zurich, Switzerland*

DYLAN J. GUERIN • *School of Life Sciences and Nevada Institute of Personalized Medicine, University of Nevada, Las Vegas, Las Vegas, NV, USA*

LEILA JAMAL • *Ruiz Department of Ophthalmology and Visual Science, McGovern Medical School at The University of Texas Health Science Center at Houston (UTHealth), Houston, TX, USA; Biological Sciences (Neurobiology, Physiology, and Behavior), College of Liberal Arts and Sciences at Arizona State University, Tempe, AZ, USA*

NANGE JIN • *Ruiz Department of Ophthalmology and Visual Science, McGovern Medical School at The University of Texas Health Science Center at Houston (UTHealth), Houston, TX, USA*

CINDY X. KHA • *School of Life Sciences and Nevada Institute of Personalized Medicine, University of Nevada, Las Vegas, Las Vegas, NV, USA*

SANGBAE KIM • *Human Genome Sequencing Center, Baylor College of Medicine, Houston, TX, USA*
TAKAE KIYAMA • *Ruiz Department of Ophthalmology and Visual Science, McGovern Medical School at The University of Texas Health Science Center at Houston (UTHealth), Houston, TX, USA*
MENG-JUNG LEE • *Natural and Medical Sciences Institute, University of Tübingen, Reutlingen, Germany; Multi Channel Systems GmbH, Reutlingen, Germany; Graduate Training Centre of Neuroscience, University of Tübingen, Tübingen, Germany*
YUMEI LI • *Human Genome Sequencing Center, Baylor College of Medicine, Houston, TX, USA*
CHIN-I LIN • *Institute of Systems Neuroscience, National Tsing Hua University, Hsinchu, Taiwan*
WEI LIU • *Department of Ophthalmology and Visual Sciences, Albert Einstein College of Medicine, Bronx, NY, USA; Department of Genetics, Albert Einstein College of Medicine, Bronx, NY, USA*
YE LONG • *Ruiz Department of Ophthalmology and Visual Science, McGovern Medical School at The University of Texas Health Science Center at Houston (UTHealth), Houston, TX, USA*
KIMBERLY A. MANKIEWICZ • *Ruiz Department of Ophthalmology and Visual Science, McGovern Medical School at The University of Texas Health Science Center at Houston (UTHealth), Houston, TX, USA*
CHAI-AN MAO • *Ruiz Department of Ophthalmology and Visual Science, McGovern Medical School, The University of Texas Health Science Center at Houston (UTHealth), Houston, TX, USA*
TAKE MATSUYAMA • *Laboratory for Retinal Regeneration, Center for Biosystems Dynamics Research, RIKEN, Kobe, Japan*
JEAN-MARC MATTER • *Department of Molecular Biology, Sciences III, University of Geneva, Geneva, Switzerland; Department of Biochemistry, Sciences II, University of Geneva, Geneva, Switzerland*
CHIZU NAKAMOTO • *Department of Biology, Neils Science Center & Center for the Sciences, Valparaiso University, Valparaiso, IN, USA*
MASARU NAKAMOTO • *Department of Biology, Neils Science Center & Center for the Sciences, Valparaiso University, Valparaiso, IN, USA*
KAZUHIRO OKA • *Department of Molecular and Cellular Biology, Baylor College of Medicine, Houston, TX, USA*
ROSS A. POCHÉ • *Department of Molecular Physiology and Biophysics, Baylor College of Medicine, Houston, TX, USA; Development, Disease Models and Therapeutics Graduate Program, Baylor College of Medicine, Houston, TX, USA; Genetics and Genomics Graduate Program, Baylor College of Medicine, Houston, TX, USA*
CHRISTOPHE P. RIBELAYGA • *Ruiz Department of Ophthalmology and Visual Science, McGovern Medical School at The University of Texas Health Science Center at Houston (UTHealth), Houston, TX, USA; MD Anderson Cancer Center UTHealth Graduate School of Biomedical Sciences, The University of Texas Health Science Center, Houston, TX, USA; Program in Neuroscience, The University of Texas MD Anderson Cancer Center UTHealth Graduate School of Biomedical Sciences, Houston, TX, USA; Program in Biochemistry and Cell Biology, The University of Texas MD Anderson Cancer Center UTHealth Graduate School of Biomedical Sciences, Houston, TX, USA; Neuroscience Research Center, The University of Texas Health Science Center, Houston, TX, USA*

Tania Rodrigues • *Department of Molecular Biology, Sciences III, University of Geneva, Geneva, Switzerland; Department of Biochemistry, Sciences II, University of Geneva, Geneva, Switzerland*
Kelly Ai-Sun Tseng • *School of Life Sciences and Nevada Institute of Personalized Medicine, University of Nevada, Las Vegas, Las Vegas, NV, USA*
Hung-Ya Tu • *Laboratory for Retinal Regeneration, Center for Biosystems Dynamics Research, RIKEN, Kobe, Japan*
Xiangyun Wei • *Department of Ophthalmology, University of Pittsburgh School of Medicine, Pittsburgh, PA, USA; Department of Developmental Biology, University of Pittsburgh School of Medicine, Pittsburgh, PA, USA; Department of Microbiology and Molecular Genetics, University of Pittsburgh, School of Medicine, Pittsburgh, PA, USA*
Yi Wen • *Department of Ophthalmology, University of Pittsburgh School of Medicine, Pittsburgh, PA, USA*
Masayuki Yamashita • *Center for Basic Medical Research, International University of Health and Welfare, Ohtawara, Japan*
Zhijing Zhang • *Ruiz Department of Ophthalmology and Visual Science, McGovern Medical School at the University of Texas Health Science Center at Houston (UTHealth), Houston, TX, USA*

Chapter 1

Investigating Neurogenesis in Birds

Tania Rodrigues, Laurent Brodier, and Jean-Marc Matter

Abstract

The macula and fovea make human vision unique among mammals. An understanding of the genetic network underlying the development and maintenance of this highly specialized region is instrumental to address issues about human macula-related retinopathies. The pigeon retina, unlike currently available animal models, shares numerous key characteristics of the primate macula and represents a promising new model for the study of retinal development. We provide key elements to take advantage of this new model for the study of retina and brain development. This includes precise embryo staging, transfection of genetic material (reporter plasmid, expression vectors, siRNAs) using in ovo and ex vivo electroporation, live imaging, high-resolution confocal imaging, and data layout and instructions for data analysis.

Key words Retina, Brain, Optic tectum, Development, Avian, Pigeon, Chick, Electroporation, Live imaging, Time-lapse imaging

1 Introduction

Age-related macular degeneration (AMD) and glaucoma are the leading cause of blindness worldwide. The likelihood that early retinal developmental defects are associated with the development of such retinal neurodegenerative diseases in the adult human remains poorly understood. Meta-analysis of genome-wide association data led to the discovery of significant loci for optic disc area and vertical cup-disc ratio (VCDR) in genes involved in retinal development [1]. For instance, single-nucleotide polymorphisms (SNPs) upstream of the human atonal homolog 7 (*Atoh7*) gene—i.e., a transcriptional regulator in the production of retinal ganglion cells [2–4]—are associated with important parameters in primary open-angle glaucoma (POAG) [5]. A better understanding of how these SNPs tune gene expression and influence the robustness in retina ontogenesis remains a major challenge for the future. Achieving this understanding is very relevant as it might open new windows of opportunities to treat and even prevent the occurrence of retinal neurodegenerative diseases.

Chai-An Mao (ed.), *Retinal Development: Methods and Protocols*, Methods in Molecular Biology, vol. 2092, https://doi.org/10.1007/978-1-0716-0175-4_1,

The macula and fovea located at the center of the retina make primate visual perception unique among mammals. Analyzing gene regulation in this highly specialized region is instrumental to address issues about human retinopathies. Unfortunately, all animal models used so far to investigate retina development and eye diseases have no macula and no fovea. Nonhuman primates should provide the best animal model for advancing fundamental knowledge on visual system development and developing new treatments and cures for blinding diseases. However, ethical considerations in the use of primates in biomedical research and costs limit functional studies and sample collections throughout their full visual system development. The pigeon (*Columba livia*) retina and the human macula share a number of structural and functional properties making the pigeon a great new model system for retina development. The fact that particular developmental genetic programs are at work in the pigeon retina to allow formation of the fovea and the midget pathway is a very promising and solid groundwork to study development and aging in retinas that display the structural and functional traits typical of primate macula and fovea [6, 7].

Unlike most bird species, pigeons can reproduce every 5 weeks all year long when bred under proper light-dark cycles. Pigeons are monogamous and typically mate for life. They reach sexual maturity at the age of 7–12 months. In captivity, they live up to 15 years. About 10 days after mating, the females lay usually two eggs in a time window of 24–48 h. Both parents will incubate the eggs for 18 days. Pigeons are altricial birds, which means that chicks require parental care over a period of ~30 days. Both female and male pigeons produce a nutrient-rich substance referred to as "pigeon milk" to feed their young.

In this chapter, we provide instructions for embryonic staging. We present protocols for conducting in vivo (in ovo) and ex vivo electroporation as a way to introduce foreign DNA and siRNAs in retinal and optic tectal cells. We describe techniques for time-lapse and confocal imaging. We outline strategies to be considered for data processing.

2 Materials

2.1 Dissection and Staging of Embryos

1. Egg incubator at 38 °C; air must be kept humid (*see* **Note 1**).
2. Fertilized and incubated eggs.
3. 70% Ethanol.
4. Binocular dissecting microscope and illumination.
5. Dissecting tools: forceps, micro-spoons, scissors, etc.
6. Sterile phosphate-buffered saline (PBS).
7. 10 cm Petri dishes.
8. Millimeter paper for staging.

2.2 Ex Vivo Electroporation

1. Binocular dissecting microscope and illumination.
2. Dissecting tools: forceps, micro-spoons, scissors, etc.
3. Dissected retina.
4. Square wave electroporator (ECM 830, BTX).
5. Electroporation cuvettes, 4 mm gap (BTX or cell projects).
6. Genetic material to be electroporated: DNA plasmids, siRNA, etc.
7. Gel-loading pipette tips.
8. Sterile PBS.
9. Complete culture medium: DMEM (with phenol red) supplemented with 10% fetal bovine serum (FBS), 1% penicillin–streptomycin.
10. 3 cm Petri dishes.
11. Four-well culture dishes (Nunc).
12. Cell incubator at 37 °C and 5% CO_2.

2.3 In Ovo Electroporation

1. Egg incubator at 38 °C; air must be kept humid (*see* **Note 1**).
2. Fertilized and incubated eggs.
3. 70% Ethanol.
4. Small flashlight for candling eggs.
5. Binocular dissecting microscope and illumination.
6. Square wave electroporator (ECM 830, BTX).
7. Electroporation Tweezertrodes (BTX).
8. Dissecting tools: forceps, micro-spoons, scissors, etc.
9. Genetic material to be electroporated: DNA plasmids, siRNA, etc.
10. Fast Green FCF dye.
11. Syringe and glass needle compression fitting (Hamilton).
12. Pulled glass capillaries, 1 mm diameter (World Precision Instruments, Inc.).
13. 5 mL Syringe and 21-gauge needles.
14. Tape and/or silicone medical tape (Micropore Silicone, 3M).

2.4 Live Imaging

1. Collagen stock solution: All throughout the procedure, keep collagen and solutions on ice to avoid premature collagen polymerization. Work in sterile conditions, and use autoclaved solutions. Take 100 mg of collagen powder (type V from rat tail); add 6 mL of ice-cold 16.7 mM acetic acid. Vortex and mix with magnetic stirrer overnight at 4 °C. Add 12 mL of ice-cold sterile distilled water. Vortex well. Transfer the collagen into a small beaker on ice (*see* **Note 2**). Add 6 mL of ice-cold sterile

distilled water in the original collagen container, vortex, and transfer it to the beaker. Prepare dialysis membrane: Cut approximately 15 cm of membrane; rinse three times with water and one time with 16.7 mM acetic acid (*see* **Note 3**). Empty the membrane of any remaining solution, block the bottom with sealing clip, and immerse two-third in 4.2 mM ice-cold acetic acid solution in a beaker. Transfer collagen into the membrane; apply sealing clip approximately 3 cm above collagen level. Adjust membrane position so that the top sealing clip is immerged in acetic acid solution (*see* **Note 4**). Use magnetic stirrer overnight at 4 °C. Collect the collagen solution in a beaker, and aliquot in Eppendorf-type tubes on ice. Store at 4 °C (*see* **Note 5**).

2. Tridimensional collagen matrix (work solution): all solutions must be kept on ice to prevent premature polymerization. Mix 100 μL of sodium bicarbonate 0.28 M, 100 μL of 10× DMEM (without phenol red), 80 μL of 4.2 mM acetic acid, and 550 μL of collagen stock solution (*see* **Note 6**).
3. Complete culture medium: prepare DMEM (without phenol red) supplemented with 10% FBS and 1% penicillin–streptomycin.
4. Glass bottom dish: e.g., Willco wells.
5. Cell incubator: temperature at 37 °C and CO_2 at 5%. Used for tissue culture and collagen polymerization.
6. Widefield fluorescence microscope: microscope must be inverted and equipped for fluorescence and have a system to maintain constant 37 °C and 5% CO_2. Motorized microscope is required for multiple positions and Z-stack acquisition. Alternatively, a spinning disc confocal microscope can be used. An objective with high numerical aperture allows more light to pass through and requires lower illumination settings, thus reducing photobleaching. We recommend using a 40× objective.
7. Computer: with enough processing power for deconvolution and data processing.

2.5 Confocal Microscopy

1. Paraformaldehyde 4%: mix 8 g of paraformaldehyde powder in 200 mL PBS. Heat in water bath until solution reaches 80 °C. Do not to exceed 80 °C. Keep heating until solution clarifies (*see* **Note 7**). Filter using a funnel with Whatman paper. Cool solution on ice, aliquot, and store at −20 °C. Avoid repeated freezing and thawing. Solution is stable for 1 week at 4 °C.

2. Cavity microscope slides, coverslips, and nail polish.
3. DABCO mounting medium: prepare 428 mM DABCO solution in 43.5% glycerol in PBS. Agitate until powder is completely solubilized. Aliquot, and store at 4 °C for 1 month. Protect from light.
4. Laser scanning confocal microscope.

3 Methods

3.1 *Ex Vivo* Electroporation

When studying the development of a tissue, being able to identify cells that express a gene of interest in time and space is necessary. Electroporation is a powerful technique allowing transfer of genetic material into cells by inducing transient pore formation at the plasma membrane. The genetic material can be used to label cells (reporter vectors) or to induce variations in the level of expression of a gene (expression vector, siRNAs). Electroporation only works in the tissue directly in contact with the medium; therefore, it is necessary to remove the tissue surrounding the retina.

3.1.1 Embryo Staging

1. After incubation for the desired number of days, wipe the eggs with 70% ethanol.
2. Crack the egg, and transfer the embryo in a 10 cm petri dish filled with PBS.
3. With forceps, remove the amnion, the membrane that surrounds the embryo.
4. Place a piece of millimeter paper under the petri dish, and measure the diameter of the eye, the length of the beak, and the length of the third toe. To properly stage the pigeon embryo, refer to Table 1 and accompanying Fig. 1. Staging of chick embryonic development was published by Hamburger and Hamilton [8].

3.1.2 Retina Dissection

1. With forceps, pinch around the eye until it detaches from the embryo.
2. With forceps, peel off the retinal pigment epithelium (RPE, the gray-brown layer of tissue) from the retina (*see* **Note 8**).

3.1.3 Electroporation Using Cuvettes

1. Prepare the genetic material for electroporation. We use reporter and expression vectors in the 0.1–0.5 μg/μL range and siRNA in the 0.1 μg/μL range, diluted in PBS.
2. Fill an electroporation cuvette with enough electroporation mix to cover the retina. One hundred μL of mix is usually enough (*see* **Note 9**).
3. Put the retina in the cuvette using a weighing micro-spoon.

Table 1
Morphometric measurements for pigeon embryo staging

Embryonic day	Eye diameter (mm)	Lens diameter (mm)	Third toe length (mm)[a; b]	Beak length (mm)[c; d]
E3	0.5	n/a	n/a	n/a
E4	0.7	0.3	n/a	n/a
E5	1	0.5	n/a	n/a
E6	2	0.6	n/a	n/a
E7	3	0.8	n/a	n/a
E8	4	1	n/a	n/a
E9	5	1.2	n/a	n/a
E10	6	1.5	n/a	n/a
E11	6.5	1.8	4.5; 3	4; 2.5
E12	7	1.8–2	6; 4	5.5; 3
E13	7.5	2	6; 4	6; 3
E14	8	2	7; 5	7; 4
E15	8.5	2	8; 6	8; 4
E16	9	2.5	10; 7	8; 4
E17	9.5	2.5	10; 7	9; 4
E18-P0	10	3	12; 9	10; 5

a; b and c; d: see accompanying Fig. 1.
n/a not available

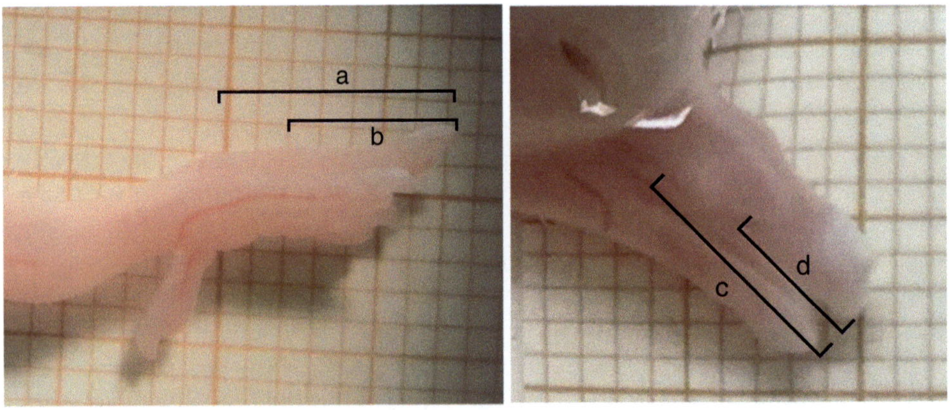

Fig. 1 Morphometric measurements for pigeon embryo staging. An example of an E14 pigeon is shown. (**a**) Third toe length from first articulation. (**b**) Third toe length from second articulation. (**c**) Beak length from proximal edge of nostril. (**d**) Beak length from distal edge of nostril

4. Using a gel-loading pipette tip, position the retina so that the lens faces toward the bottom of the cuvette. Keep in mind that the electroporated areas are parallel to the electrodes, and avoid placing the optic fissure in this area unless wanted.
5. Electroporation settings must be optimized to your needs. We most often use the following settings: five 12.5 V/cm pulses of 50 ms duration spaced 1 s apart, in both directions (*see* **Note 10**).
6. Check for bubbles on the electrodes of the cuvette: this indicates that the electric pulses happened.
7. To recover the retina, fill the cuvette with PBS, and pour the content in a 3 cm petri dish.
8. Incubate the retina in complete culture medium. We use four-well dish with 800 μL of complete culture medium.
9. The retinas are incubated at 37 °C in a 5% CO_2 cell incubator.
10. After the desired incubation time, the retina can be processed for confocal or live imaging.

3.2 In Ovo Electroporation

Ex vivo electroporation is a great and quite easy technique, but it still has some limitations. Some tissues do not survive long enough as floating explants, as is the case with the optic tectum. For some experiments, you may want to change gene expression during development and be able to check the effect in adult tissues. In this case, the in ovo electroporation procedure is required but comes with a number of challenges. Not all the embryos electroporated with this technique will survive, and the survival rate can differ greatly depending on the tissue of interest. This procedure requires a lot more patience and trial and error than the ex vivo electroporation.

3.2.1 Egg Preparation

1. After incubation for the desired number of days, 1 h prior to the electroporation, rotate the eggs for half a turn so that the embryos that were sitting on top are now at the bottom of the eggs.
2. After 30–60 min, candle the eggs, and select those where the embryo is back on top. This indicates that the membranes are not sticking to the shell, what makes the procedure easier.
3. Wipe the eggs with 70% ethanol.
4. Candle the eggs to identify the position of the embryo, and mark the shell with pencil. Also mark the position of the air chamber (*see* **Note 11**).
5. Place a piece of transparent tape at the embryo position.
6. Carefully pierce a small hole with forceps at the position of the air chamber. This allows the embryo to slightly sink into the

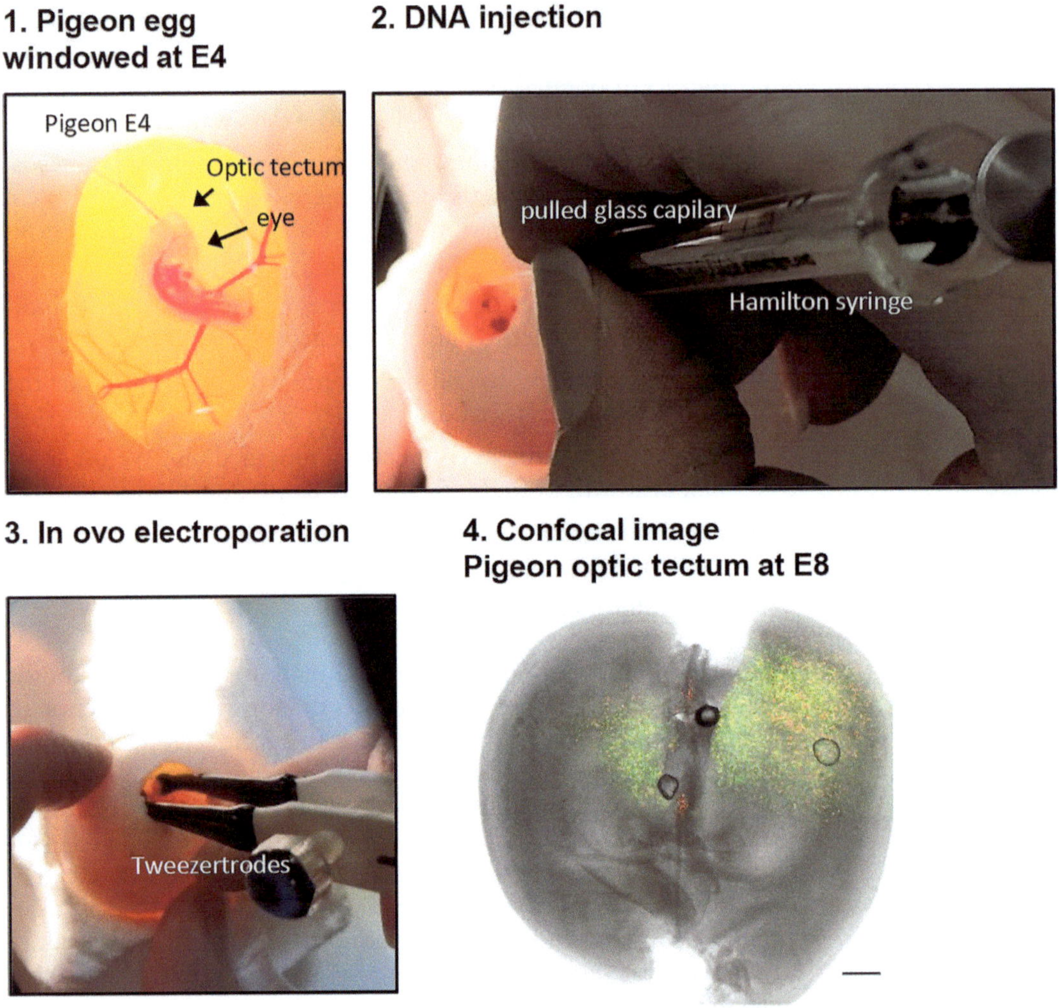

Fig. 2 In ovo electroporation of a pigeon embryo at E4. After 4 days of incubation, a window is opened in the eggshell to get access to the embryo (*1*). A Hamilton syringe with a pulled glass capillary is used to inject the DNA solution (*2*). The Tweezertrodes are placed, so the tissue of interest lies between the electrodes (*3*). The eggs are placed back in the incubator for the desired amount of time. The tissue is then dissected and fixed prior to confocal imaging (*4*). Scale bar: 500 μm

egg and allows cutting of the shell without damaging the embryo or the surrounding membranes and blood vessels (*see* **Note 12**).

7. Once the embryo is away from the shell, using very thin and small scissors, cut a window in the shell to get access to the embryo (Fig. 2 panel 1) (*see* **Note 13**).

3.2.2 Electroporation Using Tweezertrodes

1. Prepare the DNA solution to be injected. We usually inject a plasmid mix with DNA concentration in the range of 1 μg/μL and fast green dye at a final concentration of 0.025% to allow visualization of the injection.

2. Using a Hamilton syringe with a glass capillary, inject the genetic material in the tissue of interest. For instance, to electroporate the optic tectum, we inject the DNA solution in the ventricle (Fig. 2 panel 2) (*see* **Note 14**).
3. Carefully place the Tweezertrodes on each side of the tissue, and electroporate. Electroporation settings must be optimized to your needs, but we most often use the following settings: five 12.5 V/cm pulses of 50 ms duration spaced 1 s apart. Pay attention to the orientation of the electrodes for proper electroporation direction (Fig. 2 panel 3).
4. Check for bubbles close to the electrodes (*see* **Note 15**).
5. Close the windows using tape, and return the eggs to the incubator until reaching the desired stage (*see* **Note 16**).

3.3 Live Imaging

Following tissue growth over an extensive period is essential for developmental studies. Live imaging (also called real-time imaging or time-lapse imaging) of a living tissue or organism is an essential tool to understand development at cellular level [10].

The embryonic retina is convenient for live-imaging studies. Its development is largely independent from surrounding tissues, and cultured explants develop very similarly to in vivo retinas. The six types of retinal cells can be identified with genetic markers. Finally, progenitors undergo interkinetic nuclear migration (INM), i.e., back and forth movements of the nucleus along the apicobasal axis of the epithelium with mitosis usually occurring apically. These features facilitate following one cell during cell cycle progression from mitosis to mitosis.

3.3.1 Time Series Acquisition

Acquisition of time series requires immobilization of the tissue. Following markers transfection (*see* Subheading 3.1), embed retinal explants into a collagen matrix that supports the tissue and holds it in place for live imaging (Fig. 3 panel 1). We recommend glass bottom dishes with a 3 cm diameter (*see* **Note 17**). Tissue orientation relative to the microscope focal plane must be chosen according to the region and the process to study (Fig. 3 panel 2; *see* **Note 18**). Polymerize collagen 30 min at 37 °C. Add 2 mL of culture medium on top of collagen layer (Fig. 3 panel 1). Culture medium will reach the tissue by diffusion. Keep in culture for enough time (8–24 h) to allow plasmid expression before imaging.

Preheat microscope control box at 37 °C (*see* **Note 19**). Place the tissue under the microscope. Turn on CO_2 flux at 5%. Under optimal conditions, retina explants can be kept up to 72 h for live imaging.

Choose positions, and define Z-stack for acquisition. Typically, acquisition of a 40–60 μm volume in Z-stack with 1 μm intervals gives good results (*see* **Note 20**).

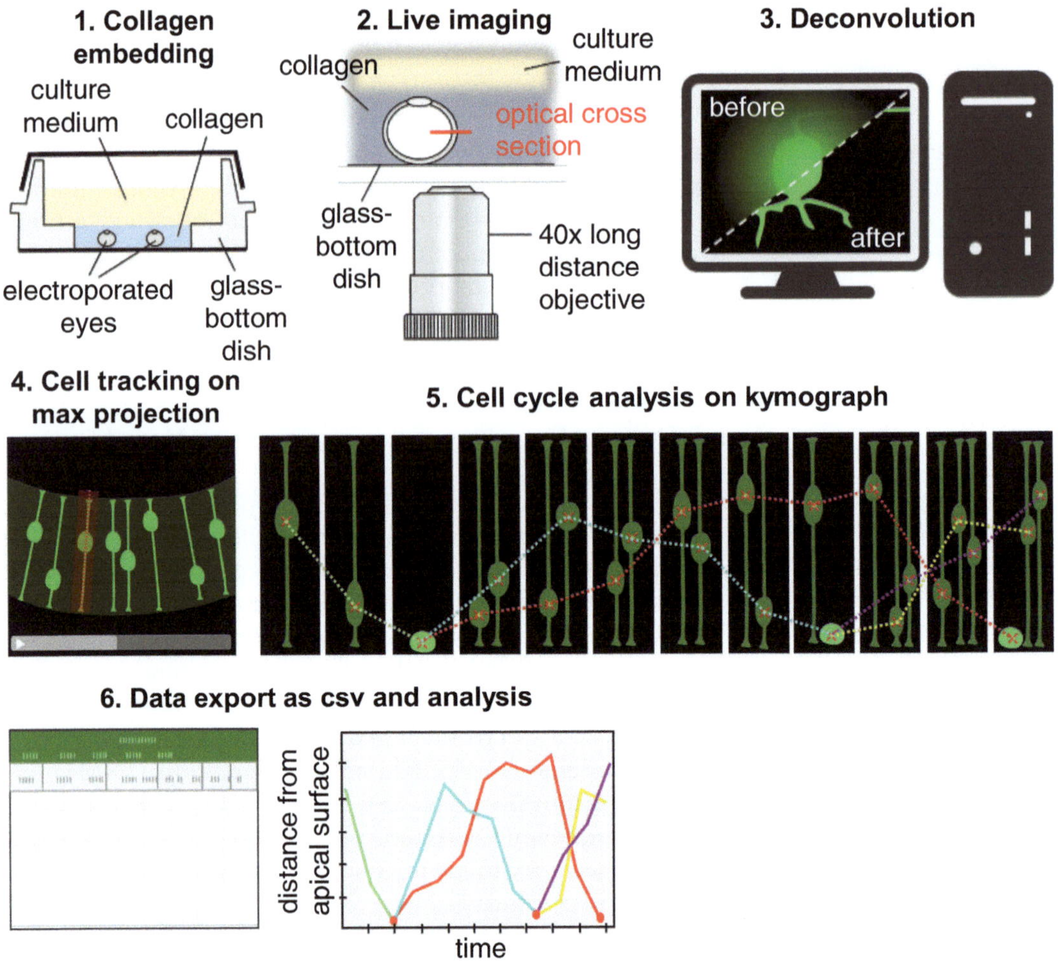

Fig. 3 Live imaging and data analysis workflow [10]. (*1*) Embed retinas in a 3D collagen matrix, and add culture medium. (*2*) Proceed to live imaging: an optical cross section at the retina equator enables to follow INM and cell cycle progression. (*3*) Improve image contrast with a deconvolution algorithm to remove the out-of-focus background noise. (*4*) Track a single cell that goes from mitosis to mitosis in a max projection of the dataset. (*5*) Generate a kymograph representation, and track mitosis as well as nucleus position on the apicobasal axis. (*6*) Export your data as .csv and proceed to analysis

Define time interval between acquisitions. To follow the cell cycle progression, we found 20 min to be the best choice since it allows detection of all cell division events. Start acquisition (*see* **Note 21**).

A variation of this method is used for tracking of retinal ganglion cell (RGC) axogenesis. Flatten retinal explants before polymerization of the collagen matrix to position RGC axons growing on the basal surface of the retina parallel to the microscope focal plane: Remove the lens and vitreous body using forceps, and inject collagen inside the retinal cup. Let a thin first layer polymerize for

30 min at 37 °C (*see* **Note 22**). Top with a second layer of collagen. Let it polymerize 30 min at 37 °C, and add 2 mL of culture of medium. Proceed with live imaging (*see* **Note 23**).

3.3.2 Dataset Processing

Live-imaging data consists of time series, typically from multiple positions, each with multiple fluorescence channels and a Z-stack.

Save your data on an appropriate informatics support (*see* **Note 24**).

Process images with a deconvolution algorithm to improve contrast and facilitate cell tracking (Fig. 3 panel 3; *see* **Note 25**).

Deconvolution creates a second dataset of the same size as the original, thereby doubling the amount of space required for data storage (*see* **Note 26**). Deconvolution involves intensive calculation and require high-processing power (*see* **Note 27**).

3.3.3 Analysis of Live-Imaging Data

Analysis of the live-imaging data can be automated to some extent. However, as no automatic analysis algorithm is prefect, we privilege semiautomatic methods that enable users to have real-time feedback over the analysis to correct errors (*see* **Note 28**).

Choose a software for image analysis. Most analysis can be carried out using the free software ImageJ or its updated distribution Fiji (*see* **Note 29**) [9]. Imaris (Bitplane) enables fast tridimensional visualization of a Z-stack and time series and contains useful tools for tridimensional segmentation and cell tracking.

The following examples illustrate analysis done in our lab from time-lapse imaging in the chick and pigeon embryonic retina.

Application 1: Assessing Cell Cycle Length and Interkinetic Nuclear Migration (INM) of Progenitors in the Retina

To analyze nuclear movements along the apicobasal axis, we use a combination of tools and plugins in ImageJ.

Look for cells that go through at least two mitoses in the live-imaging dataset. Mitosis occurs on the apical surface of the retina, and nucleus displays a very distinctive bright and round appearance (*see* **Note 30**).

Large Z-stacks include many unwanted cells that mask the cell of interest in the Z-projections. To get rid of them, perform max projections with only a few Z-slices around the cell of interest for each time point (*see* **Note 31**).

From the max projection, isolate the cell of interest. Using ImageJ "line tool," add a line from the apical to the basal surface that encompasses the cell of interest for each time point into ImageJ ROI manager. Select an appropriate line width to encompass the whole nucleus (Fig. 3 panel 4). Finally, convert lines into a series of images using the "straighten line" function of ImageJ. Combine the generated images to show all time point for a cell in a single image using the "concatenate" function of ImageJ (*see* **Note 32**).

Mark position of the nucleus in each tile by adding a "point ROI" in ImageJ ROI manager (*see* **Note 33**, Fig. 3 panel 5). Use

ImageJ "measure" function to get the *X* and *Y* coordinates of all "point ROIs" in the ROI manager (*see* **Note 34**). Convert the positional information into time and distance from the apical surface. Repeat for each daughter cells (*see* **Note 35**). Generate a .csv file for subsequent statistical analysis or plotting of INM trajectory (Fig. 3 panel 6).

Application 2: Assessing RGC Axon Growth

To track axon growth, choose preferably an axon growing outside of the electroporated region to avoid masking by surrounding fluorescent cell bodies.

Follow the position of the growth cone extremity by adding a "point ROI" in ImageJ ROI manager at each time point (*see* **Note 36**). Extract positional information at each time point with the "measure" function of ImageJ, and generate a .csv file. Convert position in pixels into distance using image scale factor, and calculate parameters (e.g., growth cone velocity, persistence of growth, duration, and frequencies of pauses) (*see* **Note 37**).

Application 3: Assessing Mitochondria Dynamics in RGC Axons

To track mitochondria dynamics in RGC axons, construct a kymograph representation: outline axon trajectory for each time point using a "segmented line" ROI in ImageJ. Choose an appropriate line width to encompass the full axon diameter. To construct the kymograph, use the "straighten line" function of ImageJ that extracts a straightened rectangular image. Keep a single horizontal array with maximal pixel intensity values along the axon. Finally, combine vertically all pixel arrays using the "concatenate" function of ImageJ (*see* **Note 38**).

In the kymograph, track single mitochondria movement over time using ImageJ ROIs. Tracks can be determined manually; however, due to the large quantity of mitochondria and time points, we recommend using an automated method (*see* **Note 39**). Calculate time and distance from positional information. Export values as a .csv file for analysis. Examples of parameters that can be calculated are velocity, persistence of movement, and proportions of mitochondria moving in the anterograde or retrograde direction and of stopped mitochondria.

3.4 High-Resolution Imaging

Analysis of tissue structure and composition with fine details requires imaging of fixed tissues. Laser scanning confocal microscopes produce high-resolution fluorescence images with minimal out-of-focus background noise. High-resolution imaging enables to resolve tissue structure with unmatched precision and is useful to look at distribution and morphology of subcellular compartments like mitochondria.

3.4.1 Confocal Imaging

After transfection with fluorescent markers, keep tissue in culture for an appropriate period that allows expression of fluorescent proteins or dye accumulation. Markers include reporter plasmids, tracker plasmids, and fluorescent dyes.

Proceed to fixation: wash the tissue with PBS. Fix the tissue explant with paraformaldehyde 4% for 20 min. Wash three times with PBS. Keep at 4 °C in PBS until mounting into microscope slide (*see* **Note 40**).

Mount fixed tissue on a cavity microscope slide with mounting medium. Cover with a coverslip, and seal with nail polish.

Proceed to microscopy: for comparison of fluorescence intensity, always use the same settings and laser intensities (Fig. 2 panel 4). Privilege quantitative mode of acquisition when available (e.g., Leica photon counting mode).

3.4.2 Analysis of Confocal Data

Laser scanning confocal imaging of the chick and pigeon embryonic retina has been used to reveal cellular organization, colocalization of fluorescent markers, and apicobasal distribution of organelles:

Analysis 1: 3D Visualization of Tissue Structure

Confocal data with a Z-stack can be visualized in tri-dimensions in Imaris (Bitplane) software, enabling analysis of tissue structure and organization as well as 3D measurements.

Analysis 2: Colocalization of Two Markers at Single Pixel Level Is Performed in Imaris (Bitplane) Software or Using the "coloc 2" Plugin Available in Fiji Distribution

Analysis of co-expression at cellular level is done in ImageJ. Define ROIs around single cells. This can be done manually or by using thresholding and segmentation functions of ImageJ. Thresholding and segmentation will give good result only if fluorescent cells are sufficiently sparse. Measure mean fluorescence value for each cell. Either convert to categorical data using thresholds (e.g., expressed/not expressed or none/low/high) or use quantitative values. Proceed to analysis. For categorical data, construct a contingency table to estimate the proportion of cells in each category. For quantitative data, construct a scatterplot, and use linear regression.

Analysis 3: Distribution Along the Apicobasal Axis

Start from a maximal projection at the equator of the retina where both apical and basal surfaces are visible. Define a line that goes from the apical to basal side of the retina. Define line width to encompass approximately 100 μm. Use the "plot profile" function of ImageJ. Values are exported to .csv for analysis.

4 Notes

1. If the incubator has no humidity control, placing a container with water at the bottom is enough.
2. We recommend doing multiple transfer of small volume with a 5 mL pipette to avoid losing collagen that will stick to the pipette wall.
3. Always keep the membrane immersed to avoid desiccation.

4. Attach membrane to the top of the beaker. You can use multiple sealing clips at the bottom to ensure complete immersion.
5. We do not recommend using collagen stock solution older than 1 year. Collagen work solution prepared from older stock solution might require adjusting pH to 7.4.
6. Collagen is highly viscous. Mix well and slowly using 1 mL pipette tip, and avoid bubbles. Pipettes designed specifically for highly viscous solution can be used.
7. The solution will remain slightly opaque, but you should see a clear difference from the fully opaque solution at the beginning.
8. Starting at the end of the optic fissure is the easiest. Never let the retina dry as this is likely to damage the cells and lead to poor electroporation efficiency.
9. It is possible to put two to three retina at the same time depending on their size.
10. We invert the cables on the electroporation apparatus to change the electroporation direction. Doing so allows you to electroporate both sides of the retina.
11. A smartphone's flashlight is enough to candle pigeon eggs as their shell is very thin, but you will need a stronger light source when candling chick eggs.
12. To help the embryos sink, you can remove some of the albumin through the hole in the air chamber using a syringe. Be careful, and orient the needle to the bottom of the eggs to avoid damages. We usually remove 1 mL when working with pigeon eggs and 2–3 mL when working with chicken eggs.
13. It is better to start with a small window and increase its size if necessary.
14. If needed, a micromanipulator can be used to hold the Hamilton syringe.
15. The electrodes of the Tweezertrodes are very small. Check for the presence of bubbles close to the electrodes in the egg.
16. Regular transparent tape can be used for chicken eggs. For the pigeon eggs, you will need some silicone medical tape. Their shell is so smooth that the regular tape will not stick for a long time.
17. Glass bottom dishes are best suited for microscopy and can hold multiple retinas in a single experiment. A dish with 3 cm diameter simultaneously minimizes the amount of collagen required to cover the tissue and allows for enough culture medium to be added.
18. An optical cross section at the retina equator is required to follow progenitor movements across the apicobasal axis. The

microscope must be equipped with a long working distance objective capable of reaching the retina equator.

19. Changes in objective's temperature affect focus position. Pre-heating ensures that the focal position remains constant.
20. Acquisition of a Z-stack enables to keep cells of interest in focus since it accounts for tissue growth that causes vertical movement in the collagen matrix.
21. Acquisition must ensure minimal stress to the tissue. Therefore, you should limit illumination intensity, frequency of image acquisition, and Z-stack to the minimum. Using a high-sensitivity camera and an objective with high numerical aperture allows using lower illumination settings.
22. The retina exhibits a tendency to wrap. Collagen depth must be kept low during the first polymerization step in order to restrain tissue orientation parallel to the dish.
23. 2–20 min interval works fine to follow axonal growth cone progression. Faster events like monitoring of axonal transport require a shorter 20 s interval. High frequency offers better temporal resolution but limits the total duration of acquisition. Typically, tissues show signs of photobleaching after 8–10 h of live imaging with a 20 s interval.
24. Depending on the number of positions and duration of the experiment, large files may be generated with datasets that can reach over 100 Gb. Consider data storage when planning live-imaging experiments.
25. Widefield imaging, unlike confocal imaging, suffers from out-of-focus fluorescence, resulting in blurry images with relatively high-fluorescence background noise. Blind deconvolution algorithms (e.g., AutoQuant by Media Cybernetics) or point spread function (PSF)-based deconvolution methods (e.g., Huygens by Scientific Volume Imaging) can be used. The second option requires generation of a PSF (experimental or theoretical) unique to each imaging setup. Note that theoretical PSF generally performs poorly compared to experimental PSF or blind deconvolution.
26. You can keep the deconvoluted dataset only during image analysis and delete it afterward, since the deconvoluted dataset can be regenerated from the original files using the same deconvolution parameters.
27. Deconvolution is a time-consuming process even on a powerful computer. Time required will largely depend on computer performances. Many deconvolution software enable batch processing of multiple datasets. Nevertheless, deconvolution of large datasets can sometime take 1 or 2 days to complete.

28. We developed a set of ImageJ plugins to facilitate live-imaging data analysis. These plugins are available upon request.
29. ImageJ is a software developed at the National Institutes of Health (NIH) that contains useful integrated functions for image analysis. In addition, many plugins are regularly proposed by the community free of charge. ImageJ plugins can be easily developed in Java using any Java IDE to combine core ImageJ functions or implement new ones.
30. In order to follow daughter cells until the next division, use both max projection and full Z-stack in ImageJ. First inspection is easier in max projection. Once a good candidate is spotted, a more precise inspection in the complete tridimensional dataset is performed.
31. We created a plugin to automate this step. Record the Z-stack position of the cell of interest at each time point by adding a "point ROI" associated with a specific Z-slice in ImageJ ROI manager. The plugin automatically performs max projection with selected number of Z-slices around the center defined by the point ROI for each time point.
32. These steps can be automated with an ImageJ plugin or macro. The final representation resembles and is adapted from kymographs that are generally used to represent movements of particles over time.
33. This step can be automated with a plugin that uses the "find maxima" function in ImageJ.
34. You must ensure that the option "bounding rectangle" is checked in the menu under "Analyze → set Measurements. . ." in order to display *X* and *Y* coordinates in the result table.
35. This process can be automated with a plugin that converts horizontal position into time and vertical position into distance from the apical surface for each "point ROI" and generates a . csv file.
36. This process can be automated in a plugin by using the "find maxima" function of ImageJ and by identifying the closest maxima in the next frame.
37. The values can be automatically extracted and converted within a plugin that generate the .csv file.
38. We made an ImageJ plugin to automatically process segmented lines stored in ROI manager into the final kymograph.
39. We made an ImageJ plugin for semiautomatic tracking that searches for the mitochondrial signal closest to an estimated position inferred from the past trajectory in the next time point. The result is stored in ImageJ ROI manager as a segmented line. Errors in trajectory can be corrected manually or

by running the algorithm again from any manually corrected intermediate point in the path.

40. Fixation should be optimized according to tissue size, thickness, and composition. It is recommended to avoid overfixation since it interferes with fluorescence. Some dyes are not retained or lose fluorescence following fixation, in which case fixation should be skipped. Generally, dyes that require a membrane potential to accumulate inside an organelle or dyes used to monitor live events cannot be used on fixed samples (e.g., TMRM, MitoTracker Green, or calcium indicators).

Acknowledgments

We are grateful to Philippe Delaunay (Pigeonneau De La Suisse Normande, Croisille, France) for the supply of pigeon eggs over the last six years. To meet our needs, the department of zootechnics at the University of Geneva (Geneva, Switzerland) opened in July 2019 a breeding facility with 50 pairs of pigeons (Columba livia). Upon request, we can provide members of the scientific community with embryonic tissue samples.

Funding: The Swiss National Science Foundation (grants 31003A-149458 and 31003A-175668), the Gelbert Foundation, and the state of Geneva support our laboratory.

References

1. Sakurada Y, Mabuchi F (2015) Advances in glaucoma genetics. Prog Brain Res 220:107–126. https://doi.org/10.1016/bs.pbr.2015.04.006
2. Matter-Sadzinski L, Puzianowska-Kuznicka M, Hernandez J, Ballivet M, Matter JM (2005) A bHLH transcriptional network regulating the specification of retinal ganglion cells. Development 132(17):3907–3921. https://doi.org/10.1242/dev.01960
3. Ghiasvand NM, Rudolph DD, Mashayekhi M, Brzezinski JAT, Goldman D, Glaser T (2011) Deletion of a remote enhancer near ATOH7 disrupts retinal neurogenesis, causing NCRNA disease. Nat Neurosci 14(5):578–586. https://doi.org/10.1038/nn.2798
4. Kanekar S, Perron M, Dorsky R, Harris WA, Jan LY, Jan YN, Vetter ML (1997) Xath5 participates in a network of bHLH genes in the developing Xenopus retina. Neuron 19(5):981–994
5. Springelkamp H, Hohn R, Mishra A, Hysi PG, Khor CC, Loomis SJ, Bailey JN, Gibson J, Thorleifsson G, Janssen SF, Luo X, Ramdas WD, Vithana E, Nongpiur ME, Montgomery GW, Xu L, Mountain JE, Gharahkhani P, Lu Y, Amin N, Karssen LC, Sim KS, van Leeuwen EM, Iglesias AI, Verhoeven VJ, Hauser MA, Loon SC, Despriet DD, Nag A, Venturini C, Sanfilippo PG, Schillert A, Kang JH, Landers J, Jonasson F, Cree AJ, van Koolwijk LM, Rivadeneira F, Souzeau E, Jonsson V, Menon G, Blue Mountains Eye Study—GWAS Group, Weinreb RN, de Jong PT, Oostra BA, Uitterlinden AG, Hofman A, Ennis S, Thorsteinsdottir U, Burdon KP, Wellcome Trust Case Control Consortium, NEIGHBORHOOD Consortium, Spector TD, Mirshahi A, Saw SM, Vingerling JR, Teo YY, Haines JL, Wolfs RC, Lemij HG, Tai ES, Jansonius NM, Jonas JB, Cheng CY, Aung T, Viswanathan AC, Klaver CC, Craig JE, Macgregor S, Mackey DA, Lotery AJ, Stefansson K, Bergen AA, Young TL, Wiggs JL, Pfeiffer N, Wong TY, Pasquale LR, Hewitt AW, van Duijn CM, Hammond CJ (2014) Meta-analysis of genome-wide association

studies identifies novel loci that influence cupping and the glaucomatous process. Nat Commun 5:4883. https://doi.org/10.1038/ncomms5883

6. Querubin A, Lee HR, Provis JM, O'Brien KM (2009) Photoreceptor and ganglion cell topographies correlate with information convergence and high acuity regions in the adult pigeon (Columba livia) retina. J Comp Neurol 517(5):711–722. https://doi.org/10.1002/cne.22178
7. Rodrigues T, Krawczyk M, Skowronska-Krawczyk D, Matter-Sadzinski L, Matter JM (2016) Delayed neurogenesis with respect to eye growth shapes the pigeon retina for high visual acuity. Development 143 (24):4701–4712. https://doi.org/10.1242/dev.138719
8. Hamburger V, Hamilton HL (1951) A series of normal stages in the development of the chick embryo. J Morphol 88(1):49–92
9. Schindelin J, Arganda-Carreras I, Frise E, Kaynig V, Longair M, Pietzsch T, Preibisch S, Rueden C, Saalfeld S, Schmid B, Tinevez JY, White DJ, Hartenstein V, Eliceiri K, Tomancak P, Cardona A (2012) Fiji: an open-source platform for biological-image analysis. Nat Methods 9(7):676–682. https://doi.org/10.1038/nmeth.2019
10. Florence Chiodini, Lidia Matter-Sadzinski, Tania Rodrigues, Dorota Skowronska- Krawczyk, Laurent Brodier, Olivier Schaad, Christoph Bauer, Marc Ballivet, Jean-Marc Matter (2013) A positive feedback loop between ATOH7 and a notch effector regulates cell-cycle progression and neurogenesis in the retina. Cell Rep 3(3):796–807

Chapter 2

Studying In Vivo Retinal Progenitor Cell Proliferation in *Xenopus laevis*

Cindy X. Kha, Dylan J. Guerin, and Kelly Ai-Sun Tseng

Abstract

The efficient generation and maintenance of retinal progenitor cells (RPCs) are key goals needed for developing strategies for productive eye repair. Although vertebrate eye development and retinogenesis are well characterized, the mechanisms that can initiate RPC proliferation following injury-induced regrowth and repair remain unknown. This is partly because endogenous RPC proliferation typically occurs during embryogenesis while studies of retinal regeneration have largely utilized adult (or mature) models. We found that embryos of the African clawed frog, *Xenopus laevis*, successfully regrew functional eyes after ablation. The initiation of regrowth induced a robust RPC proliferative response with a concomitant delay of the endogenous RPC differentiation program. During eye regrowth, overall embryonic development proceeded normally. Here, we provide a protocol to study regrowth-dependent RPC proliferation in vivo. This system represents a robust and low-cost strategy to rapidly define fundamental mechanisms that regulate regrowth-initiated RPC proliferation, which will facilitate progress in identifying promising strategies for productive eye repair.

Key words Eye, Retina, *Xenopus laevis*, Development, Stem cells, Regrowth, Regeneration, Neural, Retinal progenitor cells, Proliferation

1 Introduction

Studies of neural development in the African clawed frog, *Xenopus laevis*, have contributed significantly to the existing knowledge on vertebrate eye formation, including the identification of the eye-field transcription factors (EFTFs) and retinogenesis [1–8]. There are several features that make *Xenopus* a versatile system to study the eye [9, 10]. First, it is well suited for in vivo studies as *Xenopus* embryos develop rapidly and externally and can be generated in large numbers. Second, many molecular and cellular tools are available for investigating gene function [10–12]. Third, *Xenopus* embryos have relatively low culture costs as compared to

Cindy X. Kha and Dylan J. Guerin contributed equally to this work.

Chai-An Mao (ed.), *Retinal Development: Methods and Protocols*, Methods in Molecular Biology, vol. 2092,
https://doi.org/10.1007/978-1-0716-0175-4_2, © Springer Science+Business Media, LLC, part of Springer Nature 2020

mammalian models. Lastly, the mature *Xenopus* eye and the human eye have comparable structures due to the close evolutionary relationship between *Xenopus* and humans [13]. Notably, *Xenopus laevis* is also an established model for retinal and lens regeneration [14–22]. The high regenerative ability of *Xenopus laevis*, coupled with its well-understood eye developmental processes, makes it an ideal and unique platform for devising and testing strategies to promote productive eye repair.

Retinal progenitor cells (RPCs) are of strong interest because of their potential as treatment strategies for restoring visual function in the context of injury and/or disease [23]. During eye development, the multipotent RPCs derive from cells of the optic cup and generate all retinal neuron cell types and the Muller glia [24]. It is known that a number of developmental mechanisms are used during retinal regeneration [16, 25–27]. Thus, a key objective in building strategies for productive eye repair is to identify the differences and similarities between developmental and regenerative retinal progenitor cell (RPC) proliferation. However, current retinal regeneration studies are largely focused on mature eye models, making it challenging to undertake effective comparisons with developmental events, which occur in a very different context. To facilitate such studies, a developmental model of eye repair is needed. We found that *Xenopus* tailbud embryos at developmental stage (st.) 27 successfully regrew eyes after surgical ablation ([28]; Fig. 1). The regrowth process was rapid, completing within 5 days after ablation (Fig. 1a–h). The regrown eye was age-appropriate; contained the expected structures including the retina, lens, and pigmented epithelium; connected to the optic nerve; and showed visual function (Fig. 1i). Our studies also showed that eye regrowth is age-dependent, with st. 32 embryos losing this ability [29].

This new developmental model for eye repair now enables a detailed examination of how regenerative RPC proliferation can drive multi-tissue eye regrowth. Within the first 24 h after surgical ablation, there was a significant increase in proliferation in the regrowing eye but not in sham-operated eyes [28]. This result showed that productive regrowth requires multipotent RPC proliferation after injury. Moreover, regenerative RPC proliferation was specific for regrowth and not due solely to injury. Another interesting finding is that the eye regrowth observed after st. 27 ablation is not due to the retinal stem cells in the ciliary margin zone (CMZ) as these cells are only present much later at st. 35 [30]. Together, the results indicated it is most likely the embryonic RPCs that regulate eye regrowth.

Here, we provided detailed methods to study in vivo RPC proliferation in the context of *Xenopus* embryonic eye regrowth, including embryo culture, ablation surgery, and functional approaches to define cellular and molecular mechanisms that

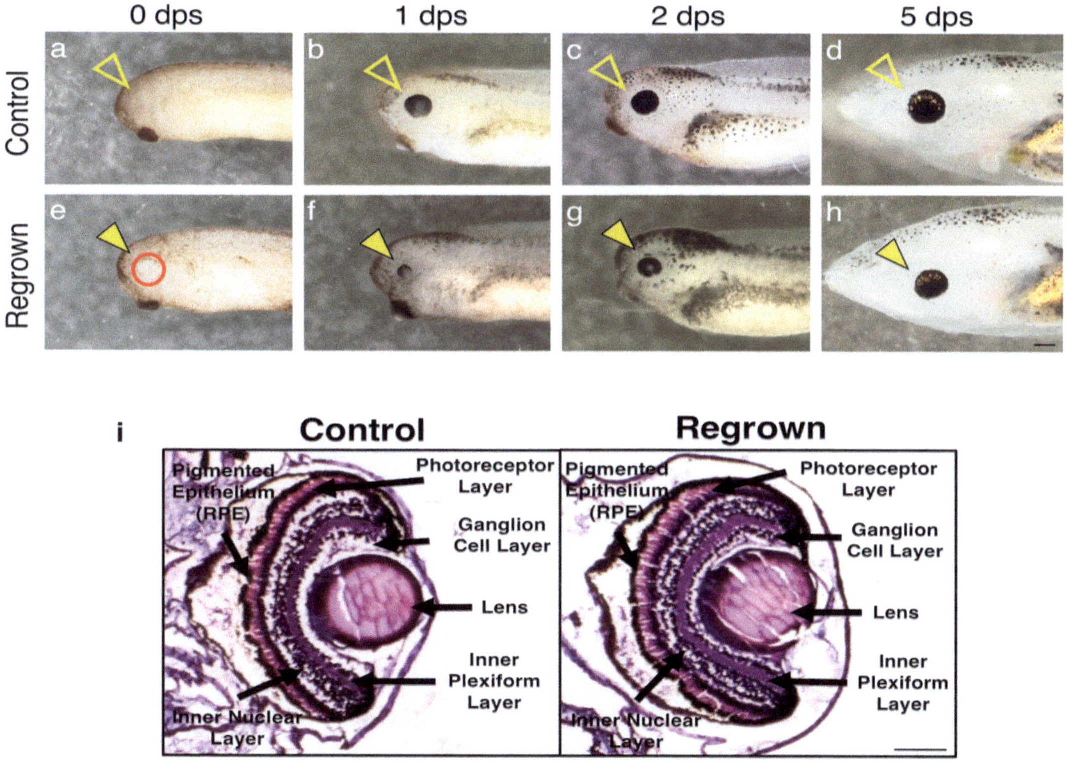

Fig. 1 Eye regrowth following surgical ablation. Images show normal eye development (**a–d**) and eye regrowth progression (**e–h**) at 0, 1, 2, and 5 days post surgery. Closed arrowheads indicate surgery site; open arrowheads indicate age-matched unoperated eye. Regrown eyes have the same eye structures compared to an unoperated sibling control. (**i**) Hematoxylin- and eosin-stained section of an unoperated sibling control (left panel) and a regrown eye at 5 days post surgery (right panel). (**a–h**) Up = dorsal, down = ventral, left = anterior, right = posterior. (**i**) Up = dorsal, down = ventral. Scale bars: (**a–h**) = 200 μm, (**i**) = 50 μm (reprinted from *Experimental Eye Research*, 169/April, 2018, Kha, C.X., Son, P.H., Lauper, J., and Tseng K.A.-S., A model for investigating developmental eye repair in *Xenopus laevis*, 38–47. Copyright 2018, with permission from Elsevier)

regulate this process. We have successfully used this model to identify apoptosis (programmed cell death) as a regeneration-specific mechanism that is required for eye regrowth [28].

In summary, the *Xenopus* developmental eye repair model described here represents a new and robust platform to interrogate in vivo retinal progenitor cell proliferation in a model vertebrate. It will enable rapid progress in distinguishing between developmental and regenerative eye mechanisms, facilitate new approaches toward stimulating RPC proliferation in vivo, and provide opportunities for translating these findings toward identifying suitable populations of stem cells for eye repair and promoting mammalian RPC proliferation in vitro and in vivo.

2 Materials

2.1 Instruments and Dissecting Tools

1. A dissecting stereo microscope
2. Two pairs of surgical forceps, No. 5 (Dumont)
3. Two pairs of AA-style forceps
4. Transfer pipets, disposable, 7.5 mL
5. Plastic Petri dishes, 60 mm × 15 mm
6. Plastic Petri dishes, 100 mm × 15 mm
7. Delicate task wipers (Kimwipes)
8. Vibratome, Leica VT1000 S or similar

2.2 General Solutions

1. 70% Ethanol in deionized water.
2. 0.1× Marc's Modified Ringer (MMR) medium: 0.1 M NaCl, 2.0 mM KCl, 1 mM $MgSO_4$, 2 mM $CaCl_2$, 5 mM HEPES, and pH 7.8.
3. 1% Agarose (Sigma-Aldrich) solution dissolved in 0.1× MMR. Heat to dissolve.
4. MEMFA fixative medium: 100 mM MOPS (pH 7.4), 2 mM EGTA, 1 mM $MgSO_4$, and 3.7% (v/v) formaldehyde [31].
5. Dejellying solution: 3% cysteine solution in deionized water and pH 7.8.
6. 4% Low-melt agarose (VWR) solution dissolved in 0.1× MMR. Heat to dissolve.

2.3 Solutions for Eye Tissue Removal Surgery

1. 5% Tricaine methanesulfonate (MS222, Sigma-Aldrich): dissolved in deionized water and stored at 4 °C.

3 Methods

3.1 Embryo Culture

1. For general *Xenopus laevis* care, induction, and fertilization of embryos, follow published protocols.
2. A protective layer of jelly surrounds the eggs. The jelly is removed from the embryos after fertilization using a 3% cysteine dejellying solution [32]. After dejellying is completed, wash the embryos several times with 0.1× MMR to completely remove the cysteine solution. Fill a 100 mm × 15 mm Petri dish with 0.1× MMR (30–35 mL). Use a transfer pipet to transfer 60–80 embryos into the dish (*see* **Note 1**).
3. Culture embryos to the desired stage. Embryos are staged using the Nieuwkoop and Faber developmental staging series [33] and can be grown in temperatures ranging from 14 to

25 °C. The rate of development is dependent on the culture temperature and embryo density. Here are general guidelines for developmental timeframe: one-cell embryos develop into st. 27 embryos in ~1.5 days at 22–25 °C, ~2 days at 18 °C, and ~3 days at 14 °C (*see* **Note 2**).

4. Monitor the growth of embryos daily. Use AA-style forceps to move and examine embryos under a stereo microscope. It is critical to maintain clean and healthy cultures. Use a transfer pipette to remove any dead embryos. Medium should remain clear. Replace cloudy medium with fresh 0.1× MMR as needed.

3.2 Preparations for Surgery

1. Set up a clean work area for surgery by wiping all surfaces with 70% ethanol, including the dissection microscope stage and surgical forceps. Spray 70% ethanol onto Kimwipes, and use the wipes to gently clean forceps tips.
2. An agarose-lined dish can be used as an aid to hold embryos in place for surgery (*see* **Note 3**). First, dissolve 1% agarose in 0.1× MMR using a microwave. Allow the solution to slightly cool before pouring the solution into a 60 mm × 15 mm Petri dish until the bottom is fully covered (~10 mL). After the agarose has solidified and cooled to room temperature, create an indentation using the tip of a 200 μL pipette tip to create a well in the agarose wide enough to hold the embryos in place. Next, fill the dish with 10–15 mL of 0.1× MMR. Add in one to two drops of 5% MS222 with a disposable transfer pipet, and then gently swirl the plate to mix to reach a final concentration of 0.02% (*see* **Note 4**).
3. Set aside two additional 60 mm × 15 mm Petri dishes for washing the animals out of the anesthetics used. Fill each dish with 10–15 mL of 0.1× MMR.
4. Fill one 100 mm × 15 mm Petri dish with 30–35 mL of 0.1× MMR to use as a culture plate for animals after surgery.

3.3 Eye Ablation Surgery

1. We have carefully studied the eye regrowth process at st. 27 tailbud embryo and observed robust retinal progenitor proliferation ([28]; *see* **Note 5**). Tailbud embryos at st. 27 can be identified by examining their external morphology as described by Nieuwkoop and Faber [33]. Our studies showed that eye regrowth ability is lost after st. 32 [29].
2. To anesthetize the embryos, use a transfer pipette to gently transfer five to ten tailbud embryos to the 1% agarose dish containing 0.02% tricaine in 0.1× MMR. The embryos should become unresponsive within a few minutes.

3. Place the dish containing embryos underneath a stereo microscope to visualize the embryos and to perform surgical procedures.
4. A transparent vitelline membrane surrounds the tailbud embryo. This membrane needs to be removed prior to surgery to allow direct access to the eyes [34]. To remove the vitelline membrane, first use a pair of No. 5 forceps to pinch the membrane in the middle posterior region of the embryo while holding the embryo in place. With a second pair of No. 5 surgical forceps, pinch the membrane at a location adjacent to the first pair. While holding the membrane with both pairs of forceps, pull the forceps in opposite directions to gently break apart the membrane and release the tailbud embryo (Fig. 2a). Allow the embryos 10–15 min to gradually straighten (Fig. 2b) out prior to beginning the next steps.

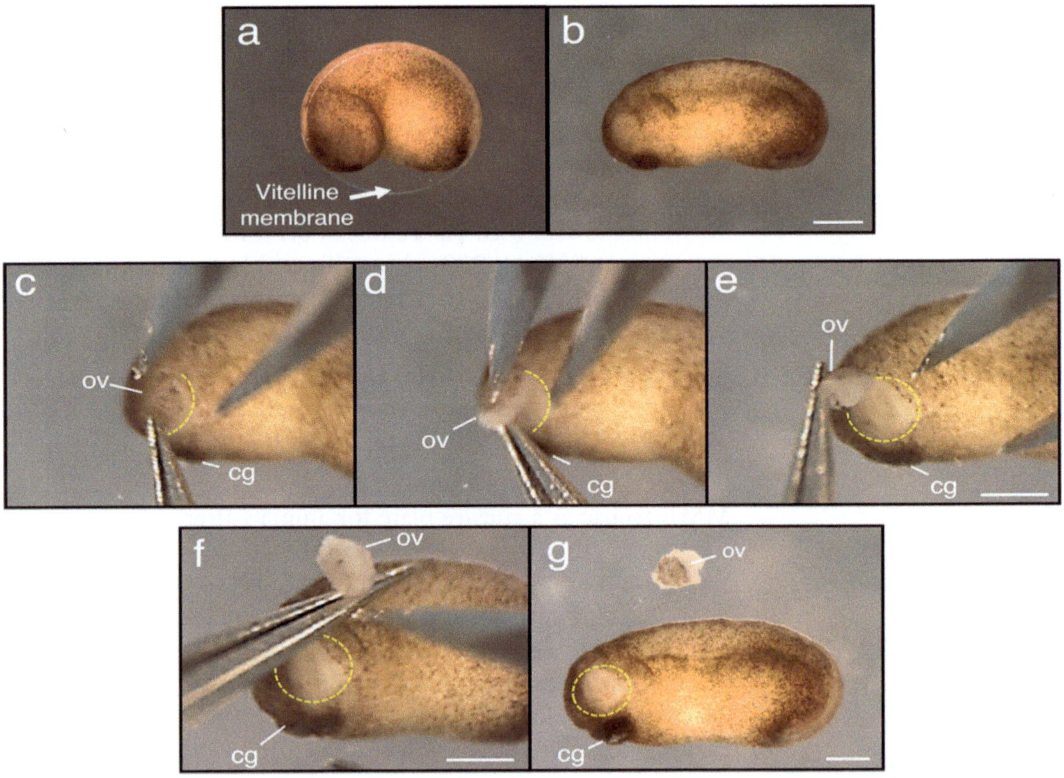

Fig. 2 Key steps in the eye ablation protocol. (**a**) The vitelline membrane (indicated by an arrow) is translucent and encases the tailbud embryo. (**b**) Removal of the vitelline membrane enables the embryo to straighten and allows access to the eye. (**c**–**g**) Images showing eye ablation surgery. (**c**) An initial cut is made using sharp forceps. (**d**) The cut is continued around the outline of the eye. (**e**–**g**) The eye tissues protrude during the surgery, and the tissues can be removed as one intact embryonic eye. *ov* optic vesicle, *cg* cement gland. Up = dorsal, down = ventral, left = anterior, right = posterior. Scale bars: (**a**–**g**) = 500 μm

5. Place the embryos into the indentation(s) made in the agarose plate. Using a pair of AA-style forceps, orient the embryos laterally with the same side (either right or left) facing upward (*see* **Note 6**).
6. In st. 27 tailbud embryos, the eye is easily identified at the head region as it protrudes out. At this stage, the embryonic eye contains the differentiating lens placode and an eye cup, with retinogenesis having started at st. 24. Use a pair of sharp No. 5 surgical forceps to make an initial surgical incision into the eye. This can be done by angling the forceps tips to make a small incision at the protruding edge of the eye (Fig. 2c). At the same time, a second pair of forceps can be used to brace the body of the animal during surgery (*see* **Note** 7).
7. After the initial cut, some eye tissues will bulge out slightly (Fig. 2c, d). Using the sharp tips of the forceps, continue cutting around the outline of the eye until the protruding tissues are completely excised and removed from the embryo (Fig. 2e–g; *see* **Note 8**).
8. After surgery, allow the embryo to recover in 0.1× MMR for 3–5 min. Remove the embryo from the tricaine solution by gently transferring the operated embryo to a Petri dish containing 0.1× MMR using a transfer pipet (*see* **Note 9**).
9. Perform a second wash by transferring the operated tailbud embryos to a second Petri dish containing 0.1× MMR. Maintain animals in 0.1× MMR at all times. It is important to minimize the amount of solution transferred between dishes to avoid transferring residual tricaine during the wash steps.
10. After the second wash, transfer the operated tailbud embryo to the culture plate. Observe embryos for normal wound closing at the surgery site (2–3 h). Culture the embryos in a 22 °C incubator for 1–5 days as needed.
11. For individual experiments, generally, 20–30 embryos are needed. Set aside a similar number of age-matched unoperated embryos to serve as controls.

3.4 Assessment of Eye Tissue Removal

1. Assessment of eye surgery can be performed using a combination of tissue sectioning and immunofluorescence microscopy. To quantify the amount of tissue removed by surgery, first fix operated embryos after surgery in MEMFA for 1–3 h at room temperature or overnight at 4 °C. Embed fixed embryos in 4% low-melt agarose, and generate sections through the eye region using a vibratome as described in [35].
2. For each tailbud embryo, generate three to four transverse sections of 50 μm thickness through the surgery site. Immunostain sections with primary antibodies to identify eye tissues (Figs. 3a and 4a, b; Table 1). The pan-neural marker, Xen1,

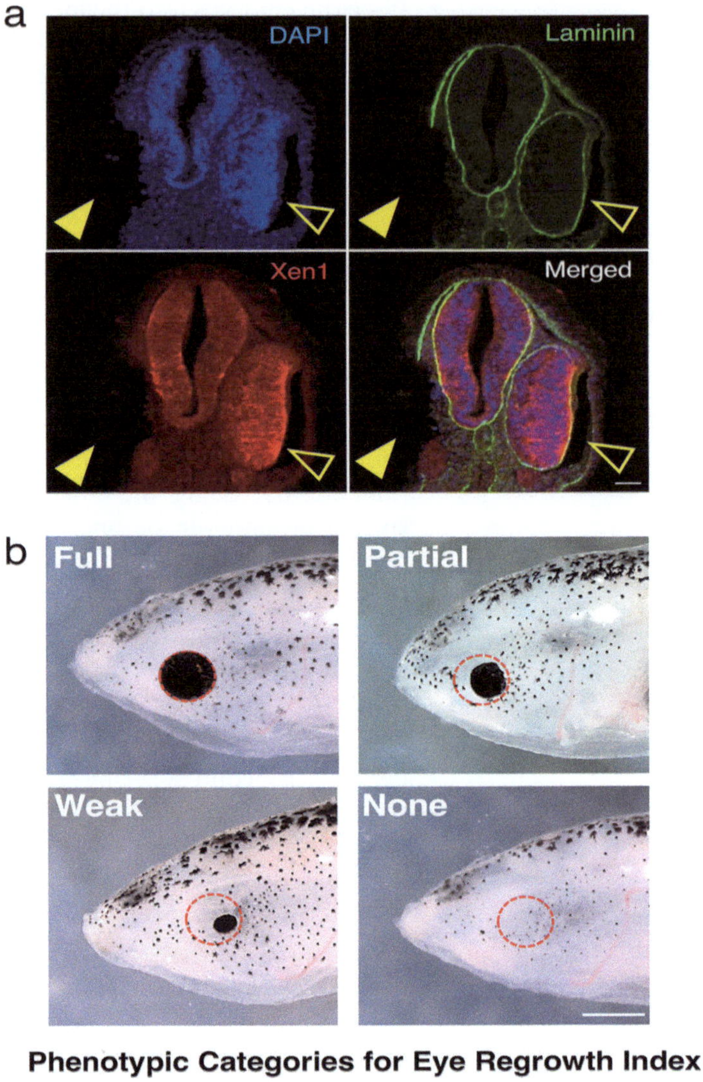

Fig. 3 Assessment of eye ablation and eye regrowth. (**a**) Shown are representative images after surgery to quantify the extent of tissue removal. Images are immunostained, transverse sections through the eye of a st. 27 tailbud embryo after surgery. Closed arrowheads indicate surgery site; open arrowheads indicate unoperated eye. Blue color indicates nuclear staining (DAPI). Green color indicates the basal lamina (anti-lamina), and outlines the optic vesicle. Red color indicates neural tissues (Xen1). (**b**) Representative images of a regrown eye following 5 days post surgery. Each regrown eye was scored based on four phenotype categories. Full = eye of appropriate size with lens. Partial = eye with minor abnormalities and comparably smaller. Weak = eye tissues with abnormal and/or absence of most eye structures. None = no visible eye tissues. (**a**) Up = dorsal, down = ventral. (**b**) Up = dorsal, down = ventral, left = anterior, right = posterior. Scale bar: (**a**) = 25 μm, (**b**) = 300 μm (reprinted from *Experimental Eye Research*, 169/April, 2018, Kha, C.X., Son, P.H., Lauper, J., and Tseng K.A.-S., A model for investigating developmental eye repair in *Xenopus laevis*, 38–47. Copyright 2018, with permission from Elsevier)

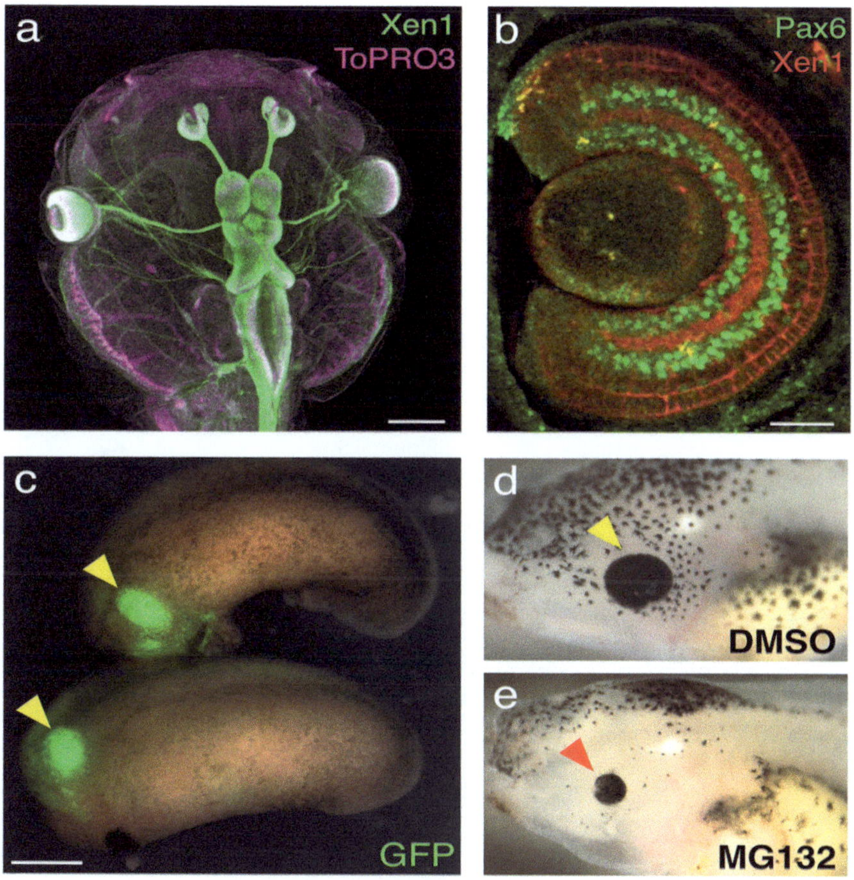

Fig. 4 Methods to study eye regrowth and RPC proliferation. (**a**) Whole-mount immunostain of a st. 46 tadpole. Green color indicates neural tissues (Xen1). Magenta color indicates nuclear signal (TO-PRO-3). (**b**) Vibratome-generated transverse eye section showing a st. 40–41 eye immunostained with anti-Pax6 and Xen1 antibodies. Green color indicates retinal cells in the ganglion and inner nuclear layers (anti-Pax6). Red color indicates neural tissues (Xen1). (**c**) Targeted microinjection of GFP mRNA into the dorsal blastomere of four-cell embryos resulted in high expression of GFP in the eye region by st. 22. (**d**, **e**) Chemical inhibitor treatment of embryos with a (**d**) DMSO-vehicle control and (**e**) MG132, a cell-permeable proteasome inhibitor. (**a**) Up = anterior, down = posterior. (**b**) Up = dorsal, down = ventral. (**c–e**) Up = dorsal, down = ventral, left = anterior, right = posterior. Scale bars: (**a** and **c**) = 500 μm and (**b**) = 50 μm

identifies neural tissues, including the eye cup [28]. The basement membrane surrounding the eye can be visualized using an anti-laminin antibody [28]. To assess the extent of the surgical ablation, obtain digital images of eye sections ([28]; *see* **Note 10**). Select the section containing the largest amount of remnant eye tissue (as labeled by the Xen1 antibody). Measure the area of the remnant eye tissue and the contralateral unoperated control individually to calculate the percentage of eye tissue ablated (Fig. 3a).

Table 1
Published antibodies for *Xenopus* eye

Target	Antigen	Source	Suggested dilution (IF)	References
Ganglion cells and inner nuclear layer cells	Anti-Islet-1, clone 39.4D5	Developmental Studies Hybridoma Bank, AB_2314683	1:200	[28, 36, 37]
Horizontal cells	Anti-PROX1	AB_37128	1:400	[38]
Müller glia, intermediate filament	Vimentin (14h7)	Developmental Studies Hybridoma Bank, AB_528507	1:50	[39, 40]
Müller glia, intermediate filament	Anti-Glutamine Synthetase	MilliporeSigma, G2781, AB_259853	1:500	[28, 36, 41]
Cone photoreceptor cells	Anti-Calbindin D-28K (EG-20)	Sigma-Aldrich, C2724, AB_258818	1:500	[28, 36, 42]
Rod photoreceptor cells	Anti-Rhodopsin, clone 4D2	EMD Millipore, MABN15, AB_10807045	1:200	[28, 43]
Rod photoreceptor cells	Anti-XAP-2, clone 5B9	Developmental Studies Hybridoma Bank, AB_528087	1:25	[40, 44]
Pax6	Anti-Pax6	BioLegend, AB_2749901	1:300	[45]
Neurofilament associated	3A10	Developmental Studies Hybridoma Bank, AB_531874	1:100	[46, 47]
Pan-neural, neural specific	Xen1, clone 3B1	Developmental Studies Hybridoma Bank, AB_531871	1:100	[28, 48]
Retinal pigmented epithelium (RPE)	Anti-Retinal Pigment Epithelium 65 (RPE-65)	MilliporeSigma, MAB5428, AB_571111	1:250	[17]
Basement membrane	Anti-Laminin	Sigma-Aldrich, L9393, AB_477163	1:300	[28, 49]
Cleaved caspase-3	Anti-Cleaved Caspase-3 (Asp175)	Cell Signaling Technology, 9661, AB_2341188	1:300	[28]
Mitosis marker	Anti-Histone H3, phospho (Ser10)	EMD Millipore, 06-570, AB_310177	1:500	[28]

3. If the eye surgery is performed correctly, the embryonic eye tissues are removed without damage to the surrounding neural and mesodermal tissues (Fig. 3a). We consistently remove ~83% of the embryonic eye tissue and observe full eye regrowth by 5 days post surgery ([28]; *see* **Note 11**).

3.5 Quantification of Eye Regrowth Quality

1. To enable the comparison of the quality of eye regrowth between different groups of embryos, a Regrowth Index (RI) was established [28] (Fig. 3b). The RI is based on four phenotype categories: (1) *Full*, a fully regrown eye with lens that is comparable in size and external morphology to an unoperated age-matched sibling; (2) *Partial*, a regrown eye with minor abnormalities and a visible reduction in eye size; (3) *Weak*, a regrown eye with no lens and severely reduced in size or a malformed regrown eye with most normal structures missing; (4) and *None*, no visible tissue regrowth of the ablated eye.
2. The phenotypic scoring for each regrown eye is normally performed at 5 days post surgery (dps), when embryos have reached the tadpole stage (the *Xenopus* eye is considered to be mature by st. 42 as it contains all the structures found in an adult eye). Anesthetize tadpoles in 0.1× MMR containing 0.02% tricaine. Examine each regrown eye, and assign the appropriate phenotype category.
3. To calculate the RI for a group of tadpoles, the following formula is used:

$$100 \times \{[3 \times (\text{number of full regrown eyes})] + [2 \times (\text{number of partial regrown eyes})] + [1 \times (\text{number of weak regrown eyes})]\}/(\text{total number of tadpoles})$$

The RI is a value ranging from 0 to 300. A value of 0 denotes no eye tissue regrowth in any individual, and a value of 300 denotes full regrowth of eye tissues in 100% of individuals in a group. Following this protocol, eye regrowth in st. 27 tailbud embryos consistently generates RI values between 280 and 290.

3.6 Molecular and Cellular Approaches to Understand RPC Proliferation

1. Retinal progenitor proliferation is required for functional eye regrowth [28]. The RI can be used to assess regrowth outcomes from loss- or gain-of-function molecular studies. We utilized this method to discover a required role for apoptosis during embryonic eye regrowth [28].
2. Operated embryos can be treated with specific chemical inhibitors dissolved in 0.1× MMR to assess the effects of inhibition on eye regrowth (Fig. 4: compare 4d (DMSO control) to 4e (treatment with MG132)) [50, 51]. Molecular inhibition using gene-specific morpholinos can be achieved with targeted injections during early embryonic stages to restrict expression of the morpholino to the eye (Fig. 4c shows GFP mRNA injection as an example) [52, 53]. When used in combination, chemical and molecular inhibition approaches represent a robust platform to identify and define mechanisms that are required for retinal progenitor proliferation in vivo during productive eye repair.

3. Gain-of-function molecular studies can be performed by microinjections of target mRNAs into embryonic cells fated to become eye tissues to induce gene overexpression (Fig. 4c).
4. To assess gene expression during eye regrowth, follow published protocols using either RNA in situ hybridization of immunohistochemistry ([54]; *see* **Note 12**).

4 Notes

1. It is common to culture *Xenopus laevis* embryos in the antibiotic gentamicin [55]. However, we have found that embryos can develop healthily without gentamicin. In this case, the embryos are monitored daily, and the culture medium is changed as needed.
2. *Xenopus laevis* developmental stage series is available online at Xenbase (http://www.xenbase.org/anatomy/alldev.do) [56]. The developmental time periods listed are approximate. Embryo density also affects developmental timing. Embryo crowding (>100 embryos per 100 × 15 mm dish) tends to delay development.
3. It is recommended to perform eye surgeries using a Petri dish lined with 1% agarose, especially for beginners. The indentations created in the agarose help to hold embryos in place and restrict movement during the procedures. Please note that the tips of No. 5 surgical forceps are sharp, delicate, and easily dented/broken, especially if they make contact with the plastic bottom of Petri dishes. The agarose plate also acts as a soft surface that protects the fine tips of surgical forceps. In general, specific care should be taken to prevent damage to the tips of the forceps. Damaged tips can be re-sharpened using sharpening tools.
4. To anesthetize embryos, use 0.01–0.03% tricaine (final concentration) in 0.1× MMR. A drop of liquid using a transfer pipet is ~50 μL. Avoid incubating embryos in tricaine for >10 min.
5. Within an individual culture plate, natural differences in growth rates will result in embryos that are in a range of developmental stages. Embryos at st. 27 can be identified by the formation of a translucent fin along the dorsal and posterior of the embryo [33]. For surgery, make sure to only select healthy tailbud embryos without developmental defects.
6. Our results indicate that there are no observable differences in the eye regrowth process between the right and left eyes (unpublished data). However, eye surgery should be performed on the same side for all embryos in an experiment to maintain experimental consistency. Anesthetized tailbud

embryos are mostly stationary. However, tailbud embryos are capable of lateral movement due to the presence of epidermal cilia on the body of the embryo. Tricaine does not inhibit ciliary movement; therefore, occasional embryonic lateral movement may occur.

7. All surgical procedures should be performed with clean surgical forceps. Remaining tissues on the forceps should be cleaned off in-between surgeries to avoid contamination.
8. The initial incision will result in a protrusion of the eye tissues. Do not dig deep into the wound site when continuing the cut around the outline of the eye cup. This may result in damage to the optic stalk and underlying brain structures.
9. Always keep the embryos submerged in the MMR solution. If the open wound of an embryo is exposed to the air-water interface, it will break open the embryo.
10. Quantitative analyses of our surgeries showed that on average, ~83% of eye tissues were consistently removed [28]. About 40% of operated embryos have less than 10% of eye tissues remaining in the embryo.
11. If embryo sections show that surrounding tissues (especially the neural tissues) are damaged during surgery, then adjust surgical excision technique by decreasing the depth of the forceps incision in the eye.
12. Xenbase (xenbase.org) contains information on commercial antibodies that have been successfully used in *Xenopus laevis*.

Acknowledgments

Funding: This work was supported by grants from the National Institute of Health (P20GM103440) and University of Nevada, Las Vegas (Faculty Opportunity Award and a Doctoral Dissertation Graduate Assistantship) to K.T.

References

1. Zuber ME, Gestri G, Viczian AS et al (2003) Specification of the vertebrate eye by a network of eye field transcription factors. Development 130:5155–5167
2. Andreazzoli M (2009) Molecular regulation of vertebrate retina cell fate. Birth Defects Res C Embryo Today 87:284–295
3. Chang WS, Harris WA (1998) Sequential genesis and determination of cone and rod photoreceptors in Xenopus. J Neurobiol 35:227–244
4. Holt CE, Bertsch TW, Ellis HM et al (1988) Cellular determination in the Xenopus retina is independent of lineage and birth date. Neuron 1:15–26
5. Wong LL, Rapaport DH (2009) Defining retinal progenitor cell competence in Xenopus laevis by clonal analysis. Development 136:1707–1715
6. Zaghloul NA, Yan B, Moody SA (2005) Step-wise specification of retinal stem cells during normal embryogenesis. Biol Cell 97:321–337
7. Viczian AS, Vignali R, Zuber ME et al (2003) XOtx5b and XOtx2 regulate photoreceptor

and bipolar fates in the Xenopus retina. Development 130:1281–1294
8. Perron M, Kanekar S, Vetter ML et al (1998) The genetic sequence of retinal development in the ciliary margin of the Xenopus eye. Dev Biol 199:185–200
9. Henry JJ, Wever JM, Natalia Vergara M et al (2008) Xenopus, an ideal vertebrate system for studies of eye development and regeneration. In: Animal models in eye research. Elsevier, Amsterdam, pp 57–92
10. Tseng A-S (2017) Seeing the future: using *Xenopus* to understand eye regeneration. Genesis 55:e23003
11. Amin NM, Tandon P, Osborne Nishimura E et al (2014) RNA-seq in the tetraploid Xenopus laevis enables genome-wide insight in a classic developmental biology model organism. Methods 66:398–409
12. Tandon P, Conlon F, Furlow JD et al (2017) Expanding the genetic toolkit in Xenopus: approaches and opportunities for human disease modeling. Dev Biol 426(2):325–335
13. Wheeler GN, Brändli AW (2009) Simple vertebrate models for chemical genetics and drug discovery screens: lessons from zebrafish and *Xenopus*. Dev Dyn 238:1287–1308
14. Day RC, Beck CW (2011) Transdifferentiation from cornea to lens in Xenopus laevis depends on BMP signalling and involves upregulation of Wnt signalling. BMC Dev Biol 11:54
15. Freeman G (1963) Lens regeneration from the cornea in Xenopus laevis. J Exp Zool 154:39–65
16. Martinez-De Luna RI, Kelly LE, El-Hodiri HM (2011) The Retinal Homeobox (Rx) gene is necessary for retinal regeneration. Dev Biol 353:10–18
17. Vergara MN, Del Rio-Tsonis K (2009) Retinal regeneration in the Xenopus laevis tadpole: a new model system. Mol Vis 15:1000–1013
18. Henry JJ, Tsonis PA (2010) Molecular and cellular aspects of amphibian lens regeneration. Prog Retin Eye Res 29:543–555
19. Underwood LW, Carruth MR, Vandecar-Ide A et al (1993) Relationship between local cell division and cell displacement during regeneration of embryonic Xenopus eye fragments. J Exp Zool 265:165–177
20. Araki M (2014) A model for retinal regeneration in *Xenopus*. In: Xenopus development. John Wiley & Sons, Inc, Oxford, pp 346–367
21. Yoshii C, Ueda Y, Okamoto M et al (2007) Neural retinal regeneration in the anuran amphibian Xenopus laevis post-metamorphosis: transdifferentiation of retinal pigmented epithelium regenerates the neural retina. Dev Biol 303:45–56
22. Martinez-De Luna RI, Zuber ME (2018) Rod-specific ablation using the nitroreductase/metronidazole system to investigate regeneration in *Xenopus*. Cold Spring Harb Protoc 2018:pdb.prot100974
23. Seiler MJ, Thomas BB, Chen Z et al (2008) BDNF-treated retinal progenitor sheets transplanted to degenerate rats: improved restoration of visual function. Exp Eye Res 86:92–104
24. Wetts R, Fraser SE (1988) Multipotent precursors can give rise to all major cell types of the frog retina. Science 239:1142–1145
25. Haynes T, Gutierrez C, Aycinena J-C et al (2007) BMP signaling mediates stem/progenitor cell-induced retina regeneration. Proc Natl Acad Sci U S A 104:20380–20385
26. Malloch EL, Perry KJ, Fukui L et al (2009) Gene expression profiles of lens regeneration and development in Xenopus laevis. Dev Dyn 238:2340–2356
27. Schaefer JJ, Oliver G, Henry JJ (1999) Conservation of gene expression during embryonic lens formation and cornea-lens transdifferentiation in Xenopus laevis. Dev Dyn 215:308–318
28. Kha CX, Son PH, Lauper J et al (2018) A model for investigating developmental eye repair in Xenopus laevis. Exp Eye Res 169:38–47
29. Kha CX, Tseng KA (2018) Developmental dependence for functional eye regrowth in Xenopus laevis. Neural Regen Res 13:1735
30. Straznicky K, Gaze RM (1971) The growth of the retina in Xenopus laevis: an autoradiographic study. J Embryol Exp Morphol 26:67–79
31. Sive HL, Grainger RM, Harland RM (2000) Early development of Xenopus laevis: a laboratory manual. Cold Spring Harbor Laboratory Press, New York
32. Sive HL, Grainger RM, Harland RM (2007) Dejellying Xenopus laevis embryos. CSH Protoc 2007:pdb.prot4731-pdb.prot4731
33. Nieuwkoop PD, Faber J (1994) Normal table of Xenopus laevis (Daudin): a systematical and chronological survey of the development from the fertilized egg till the end of metamorphosis. Garland Pub, New York
34. Sive HL, Grainger RM, Harland RM (2007) Removing the vitelline membrane from Xenopus laevis embryos. CSH Protoc 2007:pdb.prot4732
35. Blackiston D, Vandenberg LN, Levin M (2010) High-throughput Xenopus laevis

immunohistochemistry using agarose sections. Cold Spring Harb Protoc 2010:pdb.prot5532
36. Pai VP, Aw S, Shomrat T et al (2012) Transmembrane voltage potential controls embryonic eye patterning in Xenopus laevis. Development 139:313–323
37. Ericson J, Thor S, Edlund T et al (1992) Early stages of motor neuron differentiation revealed by expression of homeobox gene Islet-1. Science 256:1555–1560
38. Bertolesi GE, Hehr CL, McFarlane S (2014) Wiring the retinal circuits activated by light during early development. Neural Dev 9:3
39. Dent JA, Polson AG, Klymkowsky MW (1989) A whole-mount immunocytochemical analysis of the expression of the intermediate filament protein vimentin in Xenopus. Development 105:61–74
40. Martinez-De Luna RI, Ku RY, Aruck AM et al (2017) Müller glia reactivity follows retinal injury despite the absence of the glial fibrillary acidic protein gene in Xenopus. Dev Biol 426:219–235
41. Pahuja SL, Mullins BT, Reid TW (1985) Bovine retinal glutamine synthetase 1. Purification, characterization and immunological properties. Exp Eye Res 40:61–74
42. Gargini C, Terzibasi E, Mazzoni F et al (2007) Retinal organization in the retinal degeneration 10 (rd10) mutant mouse: a morphological and ERG study. J Comp Neurol 500:222–238
43. McIlvain VA, Knox BE (2007) Nr2e3 and Nrl can reprogram retinal precursors to the rod fate in Xenopus retina. Dev Dyn 236:1970–1979
44. Harris WA, Messersmith SL (1992) Two cellular inductions involved in photoreceptor determination in the Xenopus retina. Neuron 9:357–372
45. Rungger-Brändle E, Ripperger JA, Steiner K et al (2010) Retinal patterning by Pax6-dependent cell adhesion molecules. Dev Neurobiol 70:764–780
46. Hocking JC, Hehr CL, Chang R-Y et al (2008) TGFβ ligands promote the initiation of retinal ganglion cell dendrites in vitro and in vivo. Mol Cell Neurosci 37:247–260
47. Gordon L, Mansh M, Kinsman H et al (2010) Xenopus sonic hedgehog guides retinal axons along the optic tract. Dev Dyn 239:2921–2932
48. Ruiz i Altaba A (1992) Planar and vertical signals in the induction and patterning of the Xenopus nervous system. Development 116:67–80
49. Satoh A, Ide H, Tamura K (2005) Muscle formation in regenerating Xenopus froglet limb. Dev Dyn 233:337–346
50. Tomlinson ML, Field RA, Wheeler GN (2005) Xenopus as a model organism in developmental chemical genetic screens. Mol BioSyst 1:223
51. Tomlinson ML, Hendry AE, Wheeler GN (2012) Chemical genetics and drug discovery in Xenopus. Methods Mol Biol 917:155–166
52. Moody SA (1987) Fates of the blastomeres of the 16-cell stage Xenopus embryo. Dev Biol 119:560–578
53. Moody SA (2018) Microinjection of mRNAs and oligonucleotides. Cold Spring Harb Protoc 2018:pdb.prot097261
54. Monsoro-Burq AH (2007) A rapid protocol for whole-mount in situ hybridization on Xenopus embryos. Cold Spring Harb Protoc 2007:pdb.prot4809-pdb.prot4809
55. Elsner HA, Hönck HH, Willmann F et al (2000) Poor quality of oocytes from Xenopus laevis used in laboratory experiments: prevention by use of antiseptic surgical technique and antibiotic supplementation. Comp Med 50:206–211
56. James-Zorn C, Ponferrada V, Fisher ME et al (2018) Navigating Xenbase: an integrated Xenopus genomics and gene expression database. Methods Mol Biol 1757:251–305

Chapter 3

Three-Dimensional Culture of Mouse Eyecups

Raven Diacou, Punita Bhansali, and Wei Liu

Abstract

Retinal neurons and glia in the adult vertebrate retina are differentiated from multipotent retinal progenitors in the eyecups under the regulation of intrinsic and extrinsic factors, but the molecular mechanism underlying the process is partially understood. Functional studies using engineered mice provide tremendous insight into the mechanisms of retinal cell differentiation, but in utero embryogenesis prevents manipulations of mouse embryonic retina. Mouse eyecup culture using a culture filter or insert has been developed, but retinal structure is often altered due to the flattening of mouse eyecups in these culture systems. In this chapter, we describe three-dimensional culture of embryonic mouse eyecups. In our system, cell differentiation, stratified retinal structure, and ciliary margins in cultured eyecups were reminiscent of those in vivo. Our 3D culture of mouse eyecups has multiple applications when wild-type or engineered mice are used as models for studying retinal cell differentiation.

Key words Retinal cell differentiation, Three-dimensional culture, Mouse eyecup explant

1 Introduction

The retina in vertebrates is a stratified neural tissue that is composed of six major types of neurons and one type of glia. All cells in the adult retina are differentiated from multipotent retinal progenitors in the eyecup under the control of both intrinsic and extrinsic factors, and the order of generation of retinal neurons is evolutionarily conserved [1]. In humans, mutations in genes critical for retinal cell differentiation cause inherited retinal degenerations. Molecular dissection of retinal development is fundamental for elucidating mechanisms of inherited retinal degenerations and toward treatment of retinal diseases through regenerative medicine.

Genetically engineered mice are the leading models for studying gene regulation during retinal development and have provided tremendous insight into the mechanisms of retinal cell

Raven Diacou and Punita Bhansali contributed equally to this work.

Chai-An Mao (ed.), *Retinal Development: Methods and Protocols*, Methods in Molecular Biology, vol. 2092,
https://doi.org/10.1007/978-1-0716-0175-4_3, © Springer Science+Business Media, LLC, part of Springer Nature 2020

differentiation. However, embryogenesis in utero prevents embryonic manipulations of the retina and live imaging. To overcome these limitations, methods of retinal explant culture have been developed by several groups [2–4]. In these methods, embryonic eyecups are cultured on top of porous filters or culture inserts for several days. One drawback of filter-based methods is that the cultured retina tends to become flattened, leading to alterations of retinal structure.

In this chapter, we describe three-dimensional (3D) culture of embryonic mouse eyecups. Instead of growing eyecups on porous filters or inserts as described in other methods [2–4], we describe a method in which mouse eyecups (containing only neural retina and lens) are suspended as floating spheres in culture medium in 24-well plates on an orbital shaker. Orbital agitation efficiently prevented the attachment of eyecups to the culture surface and facilitated aeration of the culture medium. Our method does not require removal of lens in eyecups nor incisions in the neural retina to make a flower preparation [4], avoiding time-consuming procedures in dissection. We found that 3D culture of eyecups at E12.5–14.5 for several days is a practical way to study the regulation of retinal cell differentiation, because at these stages multipotent retinal progenitors are abundant and dissection of eyecups from surrounding periocular tissues is feasible. To test the effect of extrinsic factors on retinal development, candidate factors can be added to the culture medium, and the resulting eyecup cultures can be analyzed by immunostaining, in situ hybridization, live imaging, and quantitative RT-PCR. We cultured mouse eyecups at E12.5–14.5 for up to 6 days. Using the method described here, after several days of culture, cell differentiation markers were expressed, stratified structure of the retina was preserved, and ciliary margins formed in the cultured eyecups. Previous methods of retinal explant culture described the degeneration of retinal ganglion cells due to a lack of neurotrophic factors [3]. Similarly, in our 3D culture of mouse eyecups, retinal ganglion cells initially differentiated properly but subsequently degenerated after several days of culture, confirming the previous findings. Collectively, our 3D culture of mouse eyecups is useful in functional studies of extrinsic factors, gene delivery, tests of gene constructs, live imaging, and overcoming mouse lethality when engineered mice are used as models for studying retinal cell differentiation (Figs. 1 and 2).

2 Materials

2.1 Instruments

1. Dissecting microscope with dual gooseneck illumination
2. Orbital shaker
3. 37 °C Water bath

Fig. 1 A picture of homemade curved Pasteur pipet

4. Tissue culture hood
5. Tissue culture incubator 37 °C/5% CO_2
6. Inverted microscope

2.2 Dissection Tools

1. Small scissors (to cut pregnant dam abdomen)
2. Two Dumont no. 3 forceps (embryo collection)
3. Two Dumont no. 5 or no. 55 forceps (fine dissection)

2.3 Dissection

1. 70% Ethanol
2. DMEMF12 medium (Gibco)
3. Bucket with ice
4. 60 × 15 mm Petri dish
5. 18G × 1½ in Needle (BD)
6. P1000 Pipet

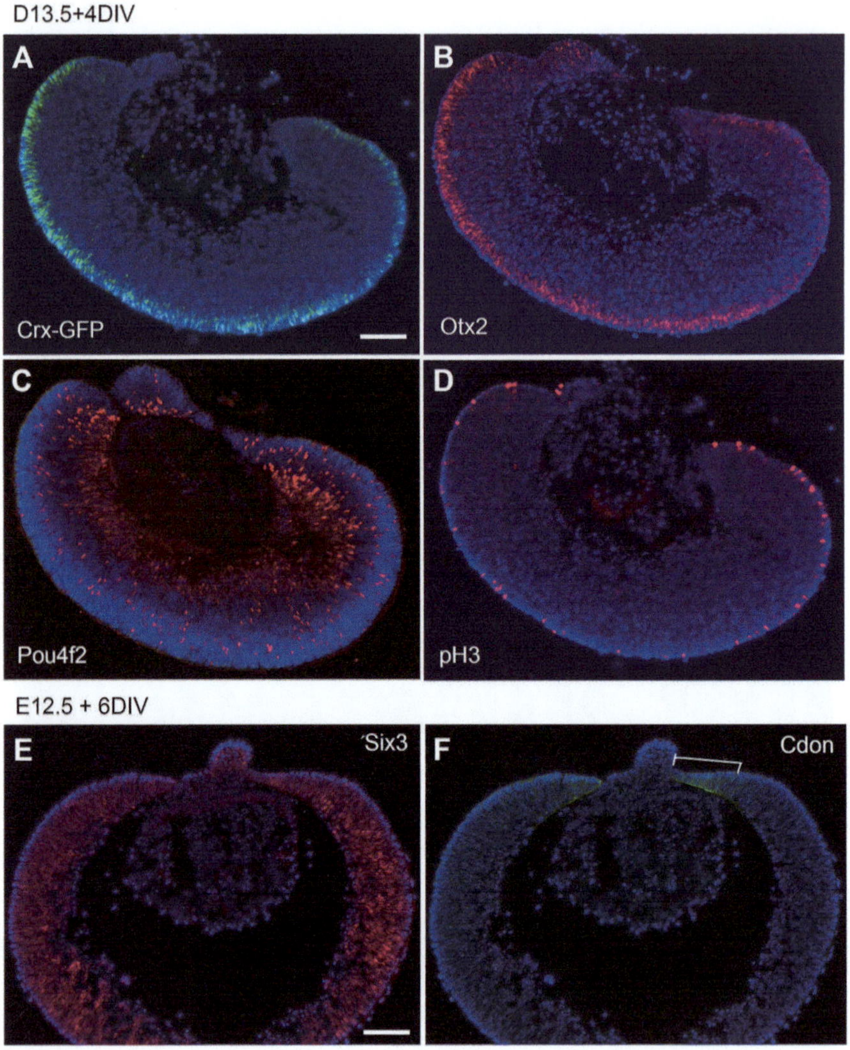

Fig. 2 Cultured mouse eyecups display proper cell differentiation, stratified retinal structure, and ciliary margins. Mouse eyecups after several days of 3D culture were analyzed using immunostaining on sections. (**a–d**) E13.5 mouse eyecups cultured for 4 days expressed photoreceptor markers Crx and Otx2, retinal ganglion cell marker Pou4f2, and proliferation marker pH3 in the appropriate layers. (**e–f**) E12.5 mouse eyecups cultured for 6 days expressed retinal progenitor marker Six3 and ciliary margin marker Cdon. Scale bars, 100 μm

2.4 Eyecup Culture

1. Culture medium (DMEM/F12 (3:1), 8% fetal calf serum (Gibco), 2% B27, 1× NEAA, 100 μM taurine (Sigma-Aldrich), and 2 mM GlutaMAX, 1× PenStrep)
2. Sterile curved Pasteur pipet with cotton plug (*see* **Note 1**) (Fig. 1)
3. Rubber bulb for glass pipet
4. Twenty-four-well plate

2.5 Cryopreservation and Embedding

1. Fresh 4% PFA/PBS (*see* **Note 2**)
2. 1× PBS
3. 30% Sucrose-1× PBS
4. OCT freezing medium (Tissue-Tek)
5. Tissue-Tek Cryomolds, 4565, Sakura Finetek USA
6. Dry ice or cryostat freezing mechanism

2.6 Immunohistochemistry

1. PAP pen (Diagnostic BioSystems)
2. 1× TBST
3. Blocking buffer (1XB)—(1/5 10% blocking buffer (Roche) in maleic acid buffer, 1/5 maleic acid buffer, 3/5 fetal calf serum)
4. Chamber for slides
5. Coplin Glass Straining Jar (Mopec)
6. VECTASHIELD Antifade Mounting Medium (Vector laboratories)

3 Methods

3.1 Embryo Retrieval

Embryo and retina retrieval should be carried out quickly to preserve retinal tissue integrity:

1. Clean all dissection tools with 70% ethanol prior to dissection. Prepare 20 mL of DMEMF12 in a 50 mL conical centrifuge tube (*see* **Note 3**).
2. Euthanize pregnant dam at E12.5 with carbon dioxide in an enclosed chamber for 7 min (*see* **Note 4**).
3. Wet the abdomen with 70% ethanol.
4. Use forceps to lift the lower abdominal region of the pregnant dam, and cut with scissors (roughly 1.5 in opening). The uterine sac should be visible as a string of embryos. Separate the embryonic sac from the uterine vasculature with scissors.
5. Place the embryos in 20 mL DMEMF12 or enough to cover the embryos.

3.2 Embryo Dissection

1. Prepare two 60 × 15 mm petri dish of ice-cold DMEMF12 each with about 10 mL of medium and an ice bucket. Carry out dissections under a dissecting microscope.
2. Place embryos in petri dish. To remove each embryo from the sac, cut a slit in uterine sac with scissors, and then, use two forceps to pinch and open the embryonic sac in an area distant from the head region (visible through the sac) to prevent damage to the eye.

3. When the embryo comes out of the sac, separate the embryo from the umbilical cord by pinching with forceps, and place the embryo into a new 50 mL conical tube with ~20 mL or enough to cover the embryos.
4. Aspirate medium, and rinse embryos twice with DMEMF12. Perform all washes on a rotating shaker. Keep embryos on ice while dissecting eyecups.

3.3 Eyecup Retrieval

1. Pre-warm culture medium (*see* recipe above) in 37 °C water bath.
2. Place an embryo into a 60 × 15 mm petri dish of ice-cold DMEMF12. (*Do not decapitate the embryos prior to dissecting the retina as it makes it easier and faster to dissect. It is not necessary to pin down the head.*)
3. Use an 18GX1/12 needle to make a superficial puncture into the ventral eye at the intersection of the cornea/sclera (technically, this area is anatomically the sclera, but the RPE is visible in pigmented embryos). This aids in the separation of the eyecup from the cornea, sclera, and RPE.
4. Use two fine forceps (55) to grab each side of the corneal slit to peel apart the tissue enough that there is an opening for the eyecup to eventually come out.
5. With one fine forceps with tips closed, gently apply a sweeping motion repeatedly over the RPE to gently isolate the eyecup from the retinal pigment epithelium. The RPE will remain attached to the embryo.
6. With fine forceps, dissect away any remaining tissue to isolate an eyecup free of extraneous tissue.
7. Cut the end of P1000 pipet tip, such that the opening is wider. Collect eyecups with a P1000 pipet (set to 200 μL), and the cut pipet tip, and transfer to eyecup to a petri dish of DMEMF12 on ice. Alternatively, a curved Pasteur pipet with rubber bulb was used to transfer the eyecups. At E12.5, the ventral fissure of the retina can help provide orientation. It appears as a faint hypopigmented indentation in the ventral region of the retina.
8. Repeat dissection on the other eye.
9. Check each eyecup to verify that they are free of extraneous tissue. The eyecup should consist of the retina and intact lens. Use fine forceps to rid eyecups of extraneous tissue. Keep eyecups on ice until it is time to plate.

3.4 Eyecup Culture (See Note 5)

1. Rinse dissected eyecups with sterile DMEM/F12 three times in a 24-well plate. Transfer the eyecups under an inverted microscope in a tissue culture hood using a curved Pasteur

pipet (*see* **Note 6**). Avoid carrying extra medium during transferring the eyecups.

2. Prepare a 24-well plate with 0.5 mL of pre-warmed culture medium at 37 °C (*see* recipe above) to wells to wash eyecups.
3. Under the inverted microscope in the tissue culture hood, transfer up to three eyecups to one well using a curved Pasteur pipet with a rubber bulb (*see* **Note 7**).
4. Check that the structural integrity of the eyecups is maintained after each transfer. Plate 0.5 mL culture medium (+signaling component inhibitors, recombinant protein of choice) to each well.
5. Transfer up to three eyecups into each control or treated well. Use a new glass pipet for each transfer.
6. Place the plate on an orbital shaker (slow gentle rotation) in a 37 °C and 5% CO_2 incubator (*see* **Note 8**).
7. For each day during the incubation period, remove the dish from the incubator, and observe the cultured eyecups under an inverted microscope to confirm that the eyecups have not become attached to the surface of the culture dish. If the eyecup attaches, shake the plate by hand to release the eyecup. If eyecup attaches, retinal structure will be altered. Daily imaging of the eyecup culture is recommended.
8. On day three, change medium by gentle aspiration with a sterile pipet, and replace with 0.5 mL of new culture medium. Continue daily observation, and ensure eyecups do not attach.
9. Depending on experiment, collect eyecups on days 3–6, image the eyecups, and prepare eyecups for cryopreservation (described below, for immunohistochemistry/in situ hybridization) or for RNA purification for qPCR analysis. Whether the eyecups can be cultured beyond 6 days has not yet been tested.

3.5 Cryopreservation and Embedding of Eyecups

1. Use freshly prepared 4% paraformaldehyde.
2. Rinse eyecups in 1 mL 1× PBS three times while shaking.
3. Fix in 1 mL of 4% paraformaldehyde/PBS for 1 h at room temperature with agitation.
4. Rinse eyecups in 1 mL of 1× PBS for three times with agitation.
5. Replace PBS with 1 mL of 30% sucrose/PBS. Place the plate on a shaker at 4 °C with gentle agitation overnight (18–24 h).
6. Use a glass pipet to transfer each eyecup to a cryomold (*see* **Note 9**). Aspirate the sucrose, and add OCT to fill the cryomold. Swirl the OCT using a P200 pipet tip, and remove any bubbles in the freezing medium.

7. Orient the eyecups lens down and optic nerve up, and note the orientation of the ventral fissure. To move the eyecup, swirl the OCT medium surrounding the eyecup with a P200 pipet tip. *Note*: It is recommended to mark the location of the eyecup with a permanent marker on the plastic of cryomolds prior to freezing, as the orientation of eyecups will be invisible after OCT freezes.
8. To freeze the retinas, apply a thin layer of OCT in the cryomold mold, and freeze the thin layer. This creates a stage for the retina to avoid embedding too close to the edge. Embed the retinas in OCT, and freeze using a cryostat deep-freezing mechanism or on a bed of dry ice, or flash freezing. Store embedded blocks in −80 °C until ready to section.
9. Collect sections (10 μm) with cryostat. *Note*: When sectioning explants, we recommend placing 4–6 sections on a slide to allow adequate space for immunohistochemistry of multiple antibodies.

3.6 Immunohistochemistry Procedure

1. For immunohistochemistry, allow sections to air-dry at least 10 min. Use a PAP Pen to circle around the sections (*see* **Note 10**).
2. Wash slides in 1× TBST 15 min in a Coplin Glass Straining Jar.
3. Block for 45 min to 3 h RT in 1XB (*see* **Note 11**).
4. Replace 1XB with primary antibody diluted in 1XB. Incubate slides in a humidified chamber at room temperature overnight.
5. Rinse slides in 1× TBST 3× 10 min while shaking. Thaw 1XB during this step.
6. Aspirate 1× TBST, and incubate slides with secondary antibody (1:200) diluted in 1XB at room temperature for 3 h.
7. Rinse slides in 1× TBST 3× for 10 min with optional DAPI stain added to the last wash. Rinse slides in ddH_2O, and then, mount with VECTASHIELD mounting medium. Slides are ready for imaging. Representative images are shown (Fig. 2).

4 Notes

1. Autoclaved Pasteur pipet was heated 2 inches from the tip using Bunsen burner to make curved Pasteur pipet.
2. Our lab prepares fresh 4% PFA diluted from a vial of 16% PFA in 1× PBS.
3. Embryos can also be collected in a petri dish.
4. Follow your institutional approved animal procedures for euthanizing mice.

5. The following steps should be carried out in a tissue culture hood. Use sterile technique and sterile filter pipet tips.
6. Since the eyecups are small, it is better to transfer them under an inverted microscope using a curved Pasteur pipet. We found that homemade curved Pasteur pipet is quite useful in transferring eyecups (Fig. 1). Otherwise, it is easy to lose eyecups or carry extra volume of medium during transferring.
7. Try to transfer minimum volume of liquid in each transfer.
8. The use of an orbital shaker in culture is important. Without the orbital agitation, mouse eyecups easily attach to culture surface and change morphology. Orbital agitation also facilitates aeration of the culture medium.
9. When embedding eyecups in a cryomold, a curved Pasteur pipet is used in transferring eyecups and removing extra liquid.
10. Do not apply the PAP pen close to the sections because it is a liquid repellant.
11. The volume depends on section area, but generally, 50–100 μL is needed per section.

Acknowledgments

We are thankful to Dr. R. Chuck for the support, grants from BrightFocus (M2012044 to W.L.), NEI (1R01EY022645 to W. L.), and RPB (unrestricted grant to the Department of Ophthalmology and Visual Science at Albert Einstein College of Medicine). R.H. and P.B. were supported by NIH training grants T32GM007288 and K12GM102779, respectively.

References

1. Cepko C (2014) Intrinsically different retinal progenitor cells produce specific types of progeny. Nat Rev Neurosci 15(9):615–627. https://doi.org/10.1038/nrn3767
2. Donovan SL, Dyer MA (2006) Preparation and square wave electroporation of retinal explant cultures. Nat Protoc 1(6):2710–2718. https://doi.org/10.1038/nprot.2006.454
3. Jin K, Xiang M (2012) In vitro explant culture and related protocols for the study of mouse retinal development. Methods Mol Biol 884:155–165. https://doi.org/10.1007/978-1-61779-848-1_10
4. Liu H, Xu S, Wang Y, Mazerolle C, Thurig S, Coles BL, Ren JC, Taketo MM, van der Kooy D, Wallace VA (2007) Ciliary margin transdifferentiation from neural retina is controlled by canonical Wnt signaling. Dev Biol 308(1):54–67. https://doi.org/10.1016/j.ydbio.2007.04.052

Chapter 4

Live Imaging of Mouse Retinal Slices

Anthony P. Barrasso and Ross A. Poché

Abstract

Live fluorescent microscopy of whole-mount rodent retinal explants has proved to be extremely valuable for understanding dynamic events during retinogenesis. However, to obtain three-dimensional images with high-quality axial resolution, investigators are restricted to specific areas of the retina and require microscopes, such as two photon, with a higher level of depth penetrance. As an alternative, we report a retinal live-imaging protocol using slice cultures that are suitable for capturing discrete cellular events during retinal development and differentiation. This is a significant improvement upon current methods, as it is more amenable to a wider array of imaging systems and does not compromise resolution of retinal cross-sectional area.

Key words Ex vivo, Slice culture, Live confocal microscopy, Developing retina, Progenitor cell proliferation, Interkinetic nuclear migration, Neuronal migration

1 Introduction

The mouse retina is an excellent model system in which to elucidate fundamental principles of neurodevelopment as well as the cellular and molecular mechanisms driving retinogenesis and disease [1–3]. Study of the developing retina has many technical advantages such as the relative ease of utilizing Cre-loxP technology, plasmid electroporation, and viral transduction to perform in vivo genetic loss-of-function, gain-of-function, and lineage-tracing experiments [4–7]. The embryonic and early postnatal mouse retina is also amenable to ex vivo culture while retaining most features of normal development [6, 8–10] and has the added advantage of allowing for time-lapse microscopy. Therefore, it is possible to monitor retinal development as it is occurring and capture specific events such as retinal progenitor cell (RPC) interkinetic nuclear migration (INM) and cell cycle progression, as well as neuron migration and differentiation [10, 11].

Currently, live imaging of the mouse retina is typically limited to explant whole mounts [9–19]. While this has been a

Chai-An Mao (ed.), *Retinal Development: Methods and Protocols*, Methods in Molecular Biology, vol. 2092,
https://doi.org/10.1007/978-1-0716-0175-4_4,

tremendously useful approach, significant technical hurdles remain. To image through the entire thickness of the explant, one must account for the attenuation of light within deeper layers. This issue is particularly problematic when researchers wish to capture events, such as INM, that occur along the radial axis. As a solution, many researchers have employed brighter and far-red-shifted fluorescent reporters and two-photon microscopy, which allow for deeper imaging with less phototoxicity. However, it is often the case that the required fluorescent reporter, such as fusion proteins, is not sufficiently bright. Furthermore, two-photon microscopy may not be available to a particular researcher, or the demand on institutionally shared equipment may make lengthy time-lapse experiments cost prohibitive or logistically impossible.

Here, we detail a straightforward method for live imaging of developing mouse retinal agarose slice cultures for at least 24 h. This approach circumvents the need for two-photon microscopy and provides excellent resolution of the entire retinal cross-sectional area using standard confocal microscopy. For example, we have previously demonstrated that events such as RPC cell cycle kinetics, INM, neuron differentiation, and mitochondrial dynamics are easily visualized [20].

2 Materials

Prepare agarose solutions and culture media under a tissue culture hood. Volumes are suitable for approximately six retinae. If more samples are necessary, adjust accordingly.

2.1 Agarose Media and Culture Media

1. DMEM/F12 without phenol red.
2. GeneMate low-melt agarose.
3. Penicillin/Streptomycin (Pen/Strep).
4. Fetal bovine serum.
5. Insulin, human recombinant, and zinc solution.
6. 50 mL conical tubes.
7. Hot plate.
8. 1 L Beaker.
9. 600 mL Beaker
10. *Agarose media recipe*: 6.5% agarose in DMEM/F12 and 1.5% agarose in DMEM/F12. *Agarose media preparation*: Pour 40 mL of DMEM/F12 media into a 50 mL conical centrifuge tube, and add 2.6 g of low-melting agarose to make a 6.5% solution. Add 30 mL of DMEM/F12 to another conical centrifuge tube, and add 0.45 g of low-melting agarose to make a 1.5% solution. Invert both to mix. Add 300 mL of water to a 1 L glass beaker, and add 500 mL of water to a 600 mL glass

beaker. Place the 600 mL beaker inside the 1 L beaker creating a double boiler. Bring water to a boil. Heat both solutions in boiling water to melt agarose. Maintain solutions at 37 °C until ready for use (*see* **Note 1**).

11. *Culture media recipe*: DMEM/F12, 10% fetal bovine serum (FBS), 1× Penicillin/Streptomycin (Pen/Strep), and insulin (5 μg/mL). Incubate at 37 °C until ready for use.

2.2 Dissection

1. Small petri dish (35 mm × 10 mm)
2. Large petri dish (100 mm × 15 mm)
3. SnuggleSafe® microwaveable heat pad (Lenric C21 Ltd)
4. Microwave
5. Zeiss Stemi 2000 microscope (Carl Zeiss, Inc.)
6. Fine scissors
7. Graefe forceps
8. Curved Graefe serrated forceps
9. Dumont no. 5 fine forceps
10. 70% EtOH (to wash dissection instruments)
11. Dissection pads

2.3 Embedding

1. Tissue-Tek® Cryomold (10 mm × 10 mm × 5 mm)
2. Plastic transfer pipette
3. Kimwipes™ delicate task wipers

2.4 Mounting, Culturing, and Sectioning

1. Vibratome (Leica Biosystems)
2. 1× phosphate-buffered saline (PBS)
3. Double-edged razor blade (Personna American Safety Razor Company)
4. LED flashlight
5. Paint brush (Dynasty, size 3)
6. Single-edge industrial razor blade
7. Plastic transfer pipette
8. Super Glue Ultra Gel Control™ (Loctite)
9. Glass bottom culture dish (MatTek Corporation, P35G-0-10-C, 35 mm dish diameter, 10 mm glass diameter)
10. P1000 pipette and tips
11. P200 pipette and tips
12. CO_2 incubator

3 Methods

3.1 Dissection, Embedding, and Mounting of the Retinae

1. Warm the heat pad in a microwave for 2.5 min.
2. Transfer 3 mL of culture media (warmed to 37 °C) to 35 mm petri dishes. Prepare three separate dishes per pair of retinae, and maintain the dishes on the heat pad.
3. Sacrifice mouse pups (postnatal days 0–5) by CO_2 inhalation followed by decapitation with scissors or isolate embryos (E15.5–E18.5) in ice-cold PBS followed by decapitation.
4. Use scissors and blunt forceps to remove the skin covering the head.
5. Using the curved, serrated Graefe forceps, gently remove both eyes from the socket, and transfer to the first 35 mm petri dish filled with warm culture media.
6. Place the 35 mm petri dish with the first pair of retinae, under a dissection microscope. Using a pair of sharp forceps, poke a small hole at the limbus, and carefully tear away the scleral and retinal pigment epithelial (RPE) layers and anterior segment from the retinae leaving behind an intact retinal cup with the lens attached.
7. Using sharp forceps, pinch the center of the lens surface. With the other sharp forceps, gently tease apart the connections between the lens and the edge of the retinal cup. Then, pull the lens and attached hyaloid vasculature away from the retinal cup. Remove any residual iris, RPE, and hyaloid vasculature from the retina (*see* **Note 2**).
8. Using a plastic transfer pipette with a cut tip, transfer the isolated retinal cup to the second dish of warm culture media, and maintain on a heat pad until all retinae are isolated and ready to embed. Before proceeding to **step 9**, perform **steps 1–8** for all retinae.
9. Use a plastic transfer pipette with a cut tip to fill a small cryomold halfway with 6.5% agarose. With a separate cut plastic transfer pipette and with the smallest media volume possible, transfer a single retinal cup to the surface of the melted 6.5% agarose in the cryomold. At this point, the retina should be slightly floating in a bubble of media on top of the melted 6.5% agarose (*see* **Note 3**).
10. Use a twisted corner of a Kimwipe to wick away residual media surrounding the retina, and slowly add more 6.5% agarose on top of the retina to fill the mold. Next, use two sharp forceps to orient the retina under the dissection microscope. Care should be taken to not rip the retinae. Ideally, the retina should maintain its cup shape and be in the middle of the mold with the ganglion cell layer facing up (*see* **Note 4**).

11. Allow the agarose to solidify at room temperature.
12. Before proceeding, perform **steps 9–11** for all retinae.

3.2 Slicing of Agarose Blocks

1. Once all of the isolated retinae are embedded, select a single agarose block, and release it from the mold onto a petri dish.
2. Using a clean razor blade, trim the edges of the block so that a slice will fit on the glass bottom dish through which it will eventually be imaged. Prior to mounting the block, orient it so that retinal cross sections can be made (*see* Fig. 1).

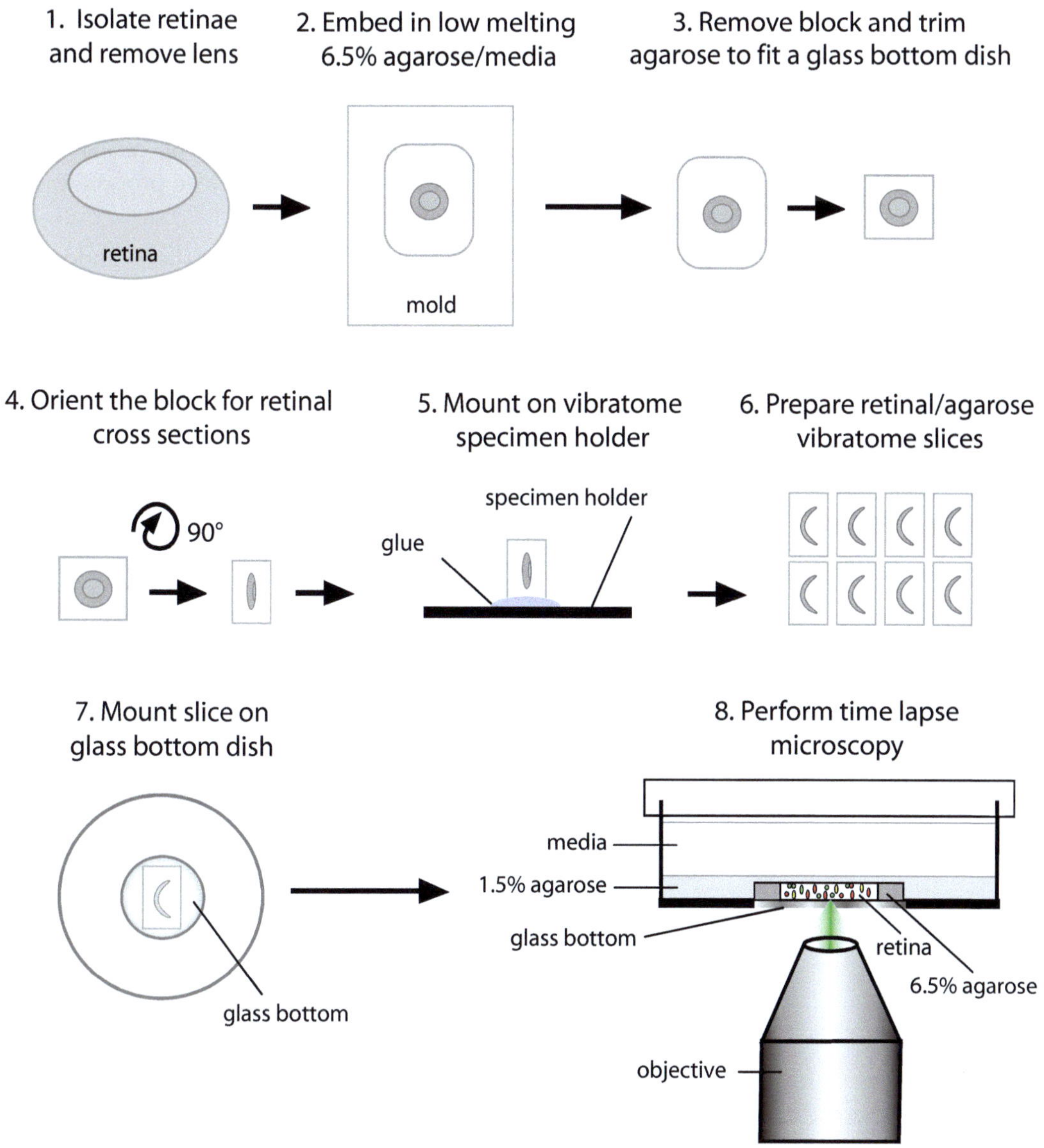

Fig. 1 Schematic of retinal slice culture protocol

3. Apply a drop of quick-dry glue to the center of the metal vibratome specimen disc. Then, using blunt forceps, immediately transfer the oriented agarose block on top of the glue drop, and gently press down. Allow the glue to dry for 5–10 min.
4. Fill the buffer tray on the vibratome with room-temperature, sterile PBS. Submerge the specimen disc with the mounted agarose block to the bottom of the buffer tray, and secure it in position. Then, secure a blade to the vibratome.
5. Begin slicing the agarose block with a frequency of 5, speed of 5, and thickness of 200 μm (*see* **Note 5**). After every pass of the knife, pay close attention to where the retina is located relative to the surface of the agarose block. Once the retina is reached, begin collecting the slices. To do so, float the slice onto the tip of a fine paint brush, and transfer the slice to the third 35 mm dish filed with warm culture media. Collect all retinal slices from the same animal in the same dish. Before proceeding to **step 6**, perform **steps 1–5** for all retinae.
6. Upon slicing all of the retinae, survey the slices under the dissection microscope. Choose the best slices (usually 1–3 per retina) that contain clear retinal cross sections in which the optic nerve head is present. These are the slices to be mounted on the glass bottom dish.

3.3 Mounting of Agarose Slices onto Glass Bottom Dishes

1. Prior to transferring the slice, pipette a drop of culture media onto the glass of the glass bottom dish creating a raised bubble. Float one of the selected retinal sections on to the paint brush, and transfer the section to the media bubble on the glass bottom dish.
2. Pipette and then wick off (with a twisted Kimwipe) as much media from the dish as possible to ensure close contact between the retinal slice and the glass.
3. Immediately overlay the agarose slice with warm 1.5% agarose media. To do so, initially cover only the slice upon the glass bottom. Once the agarose on the glass is solidified, fill the entire bottom of the dish (glass and plastic regions) with a thin layer of 1.5% agarose while also slightly covering the first agarose application. Allow this agarose to completely solidify at room temperature.
4. Overlay the solidified 1.5% agarose with liquid culture media, and immediately transfer the dish to an incubator at 37 ° C and 5% CO_2, until ready to image. Fill the dish with as much culture media as possible without overflowing the dish.
5. Repeat **steps 1–4** for all selected retinal slices (*see* **Note 6**).

3.4 Imaging

1. The microscope should be outfitted with a controlled environmental chamber kept at 37 °C and 5% CO_2 for the duration of the imaging session (*see* **Note** 7).
2. Allow the specimen to equilibrate on the heated microscope stage for at least an hour. Initially, the agarose may melt slightly, and the focus will have to be adjusted.
3. Z-stacks of the desired thickness should be acquired every 10 min for up to 48 h. We typically limit the acquisition time of a single Z-stack to 1–2 min. Occasionally, over the course of the time lapse, the sample should be checked for drift in the Z-plane and refocused if needed. Fluorescent signal intensity will vary from sample to sample, so laser power should be adjusted accordingly, but ideally not exceed 4%.

4 Notes

1. Due to autofluorescence, the DMEM/F12 media used in both the agarose and liquid culture media should not contain phenol red. Prepare more 6.5% agarose than necessary. A larger volume allows most of the air bubble to float to the surface while leaving a significant volume below that is bubble-free and better for embedding the retina. A solution of 40 mL in a 50 mL Falcon tube is usually sufficient. It is also important that FBS is not added to the culture media used to make the agarose. Upon melting the agarose, the FBS will denature, and the agarose will become cloudy making it impossible to visualize the embedded retina. FBS should be included in the liquid culture media that will overlay the agarose.
2. It is important to carefully dissect the retinae as an intact cup without poking holes or tearing the tissue. If the retina is damaged or rolls onto itself, it will be difficult to obtain sections with good cross-sectional area. From the point of retinal dissection to retinal slice culture, as much as possible, the researcher should use aseptic technique.
3. When adding the 6.5% agarose to the mold, care should be taken to avoid also transferring bubbles as they will obscure the view of the retina within the agarose block.
4. Embedding the retina in the agarose block is a critical step with several potential pitfalls. Work as quickly as possible, but do not use the agarose immediately after boiling. If the agarose does not cool to 37 °C before embedding the retina, the tissue will be damaged. Generally, after pipetting into the mold, it is best to allow the 6.5% agarose to cool at room temperature for about 1 min before embedding the retina. When removing the excess media, twist a corner of the Kimwipe into a small

tight wick, and avoid direct contact with the tissue. It is also important to be careful when positioning the retina because too much contact with forceps will damage the tissue. We find it is best to move the agarose around the tissue to create turbulent forces that aid in rotating and positioning the retina without actually touching the tissue with the forceps.

5. When sectioning with the vibratome, position the agarose block so that the first contact point with the blade is a corner rather than a flat edge. This approach helps reduce the stress on the block and tissue and reduces the chances of the retina slice falling out of the agarose slice. We found that a 200 μm slice thickness is optimal because thinner sections would fall out of the agarose and thicker sections are more difficult to image. A small LED flashlight is useful for visualizing the slices coming off of the agarose block.
6. While usually only one slice will be imaged during any given imaging session, it is best to fully prepare more than one culture. Slice quality, orientation, and fluorescent signal will vary from culture to culture. Therefore, it is best to have several options from which to choose.
7. For certain cell types and developmental processes, retinal slice culture live imaging does not necessarily require confocal microscopy or commercial environmental chambers. Mouse retinal whole-mount explants were shown to develop under CO_2-independent culture conditions [10]. Therefore, with a simply constructed homemade heater box [21] surrounding the stage of an epifluorescent microscope, one may be able to acquire adequate movies.

References

1. Bassett EA, Wallace VA (2012) Cell fate determination in the vertebrate retina. Trends Neurosci 35(9):565–573. https://doi.org/10.1016/j.tins.2012.05.004
2. Hoon M, Okawa H, Della Santina L, Wong RO (2014) Functional architecture of the retina: development and disease. Prog Retin Eye Res 42:44–84. https://doi.org/10.1016/j.preteyeres.2014.06.003
3. Veleri S, Lazar CH, Chang B, Sieving PA, Banin E, Swaroop A (2015) Biology and therapy of inherited retinal degenerative disease: insights from mouse models. Dis Model Mech 8(2):109–129. https://doi.org/10.1242/dmm.017913
4. Sanes JR (1989) Analysing cell lineage with a recombinant retrovirus. Trends Neurosci 12(1):21–28
5. Collinson JM, Hill RE, West JD (2004) Analysis of mouse eye development with chimeras and mosaics. Int J Dev Biol 48(8–9):793–804. https://doi.org/10.1387/ijdb.041885jc
6. Matsuda T, Cepko CL (2004) Electroporation and RNA interference in the rodent retina in vivo and in vitro. Proc Natl Acad Sci U S A 101(1):16–22. https://doi.org/10.1073/pnas.2235688100
7. Matsuda T, Cepko CL (2007) Controlled expression of transgenes introduced by in vivo electroporation. Proc Natl Acad Sci U S A 104(3):1027–1032. https://doi.org/10.1073/pnas.0610155104
8. Donovan SL, Dyer MA (2006) Preparation and square wave electroporation of retinal explant cultures. Nat Protoc 1(6):2710–2718. https://doi.org/10.1038/nprot.2006.454

9. Cayouette M (2003) The orientation of cell division influences cell-fate choice in the developing mammalian retina. Development 130 (11):2329–2339. https://doi.org/10.1242/dev.00446
10. Nickerson PE, Ronellenfitch KM, Csuzdi NF, Boyd JD, Howard PL, Delaney KR, Chow RL (2013) Live imaging and analysis of postnatal mouse retinal development. BMC Dev Biol 13:24
11. Huckfeldt RM, Schubert T, Morgan JL, Godinho L, Di Cristo G, Huang ZJ, Wong RO (2009) Transient neurites of retinal horizontal cells exhibit columnar tiling via homotypic interactions. Nat Neurosci 12(1):35–43. https://doi.org/10.1038/nn.2236
12. Williams PR, Morgan JL, Kerschensteiner D, Wong RO (2013) In vitro imaging of retinal whole mounts. Cold Spring Harb Protoc 2013 (1):pdb.prot072645. https://doi.org/10.1101/pdb.prot072645
13. Cayouette M, Whitmore A, Jeffery G, Raff M (2001) Asymmetric segregation of Numb in retinal development and the influence of the pigmented epithelium. J Neurosci 21 (15):5643–5651
14. Roehlecke C, Schumann U, Ader M, Knels L, Funk RHW (2011) Influence of blue light on photoreceptors in a live retinal explant system. Mol Vis 17:876–884
15. Kechad A, Jolicoeur C, Tufford A, Mattar P, Chow RWY, Harris WA, Cayouette M (2012) Numb is required for the production of terminal asymmetric cell divisions in the developing mouse retina. J Neurosci 32 (48):17197–17210. https://doi.org/10.1523/JNEUROSCI.4127-12.2012
16. Lee JE, Liang KJ, Fariss RN, Wong WT (2008) Ex vivo dynamic imaging of retinal microglia using time-lapse confocal microscopy. Invest Ophthalmol Vis Sci 49(9):4169–4176. https://doi.org/10.1167/iovs.08-2076
17. Zabel MK, Zhao L, Zhang Y, Gonzalez SR, Ma W, Wang X, Fariss RN, Wong WT (2016) Microglial phagocytosis and activation underlying photoreceptor degeneration is regulated by CX3CL1-CX3CR1 signaling in a mouse model of retinitis pigmentosa. Glia 64 (9):1479–1491
18. Liang KJ, Lee JE, Wang YD, Ma W, Fontainhas AM, Fariss RN, Wong WT (2009) Regulation of dynamic behavior of retinal microglia by CX3CR1 signaling. Invest Ophthalmol Vis Sci 50(9):4444–4451. https://doi.org/10.1167/iovs.08-3357
19. Groeger G, Mackey AM, Pettigrew CA, Bhatt L, Cotter TG (2009) Stress-induced activation of Nox contributes to cell survival signalling via production of hydrogen peroxide. J Neurochem 109(5):1544–1554. https://doi.org/10.1111/j.1471-4159.2009.06081.x
20. Barrasso AP, Wang S, Tong X, Christiansen AE, Larina IV, Poche RA (2018) Live imaging of developing mouse retinal slices. Neural Dev 13 (1):23. https://doi.org/10.1186/s13064-018-0120-y
21. Jones EA, Crotty D, Kulesa PM, Waters CW, Baron MH, Fraser SE, Dickinson ME (2002) Dynamic in vivo imaging of postimplantation mammalian embryos using whole embryo culture. Genesis 34(4):228–235. https://doi.org/10.1002/gene.10162

Chapter 5

Retinal Strip Culture for Studying Ganglion Cell Axon Growth

Masayuki Yamashita

Abstract

Retinal ganglion cell (RGC) axons extend along the inner limiting membrane, which forms the extracellular matrix (ECM) containing laminin, collagen, and proteoglycans. RGC axons express integrin, which is activated by binding to ECM proteins to regulate cytoskeleton. To study the growth of RGC axons in vitro, maintaining the natural environment for them is absolutely necessary. For this purpose, culturing a strip of embryonic chick retina in Matrigel® is a suitable method. This article describes detailed techniques of the retinal strip culture.

Key words Retinal strip, Retinal ganglion cell, Chick embryo, Axon growth, Extracellular matrix, Axon guidance

1 Introduction

Retinal ganglion cell (RGC) axons extend along the inner limiting membrane (ILM) during the early period of embryonic development. The ILM forms the extracellular matrix containing laminin and collagen [1]. Laminin in the ILM plays a pivotal role in RGC axon emergence [2], and the ILM is essential for RGC survival [3] and axon navigation [4]. For studying the development of RGC axons, maintaining the natural environment for them is absolutely necessary. However, most of previous studies on axon growth used 2-D cultures eliminating the extracellular matrix: Dissociated cells are attached to the floor of culture dishes, and neurites extend on it. In these cultures, enzymatic or mechanical dissociation process disrupts the natural environment of growing axons, and intact cell morphologies are lost; most neurites are regenerating with different morphologies from innate ones. The shape of axon tips becomes different from those growing within the original tissue; a growth cone is flattened and expands two-dimensionally on the floor of culture dishes. Such an artificial structure would never be observed in the original tissue. To maintain the natural environment for growing RGC axons in vitro, a culture method without cell

Chai-An Mao (ed.), *Retinal Development: Methods and Protocols*, Methods in Molecular Biology, vol. 2092,
https://doi.org/10.1007/978-1-0716-0175-4_5,

dissociation and providing 3-D extracellular matrix is required. For this purpose, one of suitable methods is to culture a strip of embryonic retina in a substrate similar to the ILM.

This chapter describes detailed techniques of the retinal strip culture. A retinal strip of an embryonic chick is embedded in Matrigel®, which forms the extracellular matrix containing collagen, laminin, and proteoglycans. RGC axons express integrin [5], which binds to the extracellular matrix proteins [6], and sends the intracellular signal for the linkage between the extracellular matrix and cytoskeleton [7]. Growth factors cooperate in this signaling, and the extracellular proteoglycans are essential for this cooperation [8]. By using the retinal strip culture, the effect of electric fields on the direction of RGC axon extension was successfully studied [9].

2 Materials

Prepare a clean dark room containing a small lab bench. An incubator and a dissecting microscope are put on the bench. It is convenient to set a fluorescence microscope on a vibration isolator within the dark room to take photos of axons. A conventional clean bench for working is not used.

2.1 Fertilized Eggs

1. Incubator: Put fertilized eggs in it, and set the temperature at 38.0–38.3 °C. Rotation of eggs is not necessary up to 6 days' incubation. Keep high humidity (*see* **Note 1**).
2. Refrigerator: Store fertilized eggs before the incubation for up to 2 weeks. Set the temperature at 10–12 °C. Put a dish filled with water to keep humidity.
3. Fertilized eggs: Purchase from a local farmer or a supplier of experimental animals (*see* **Note 2**).

2.2 Clean Room

1. Dark clean room equipped with a motorized ceiling fan HEPA filter unit (*see* **Note 3**).
2. Lab bench.
3. Dissecting microscope.
4. Dry incubator (37.5–37.9 °C).
5. Vibration isolator: a heavy table supported by cushions of compressed air or nitrogen gas.
6. Confocal fluorescence microscope: an upright microscope with high-magnification water immersion objectives (×20, ×40, ×100) and low-magnification objectives (×4, ×10).

2.3 Culture Chamber

1. Round culture chamber (25 mm in diameter, 2 mm in thickness) (*see* **Note 4**).

2. Cover glass (25 mm in diameter) (*see* **Note 5**).
3. Cover glass (15 mm in diameter) (*see* **Note 5**).
4. Acrylic disc (30 mm in diameter, 10 mm in thickness).
5. Six-well culture plate.
6. Airtight jar (2.5 L in volume).
7. CULTUREPAL® (Mitsubishi Gas Chemical Company, Inc., Tokyo, or Cosmo Bio) (*see* **Note 6**).

2.4 Solutions

1. Ca^{2+}-Mg^{2+}-free Hank's solution (HBSS).
2. Matrigel® (Corning) (*see* **Note** 7).
3. Fetal bovine serum (FBS).
4. Chicken serum (Gibco).
5. Serum mixture: 25 mL FBS and 5 mL chicken serum; store at 4 °C.
6. DMEM (low glucose, with phenol red).
7. DMEM (HEPES buffered, without phenol red).
8. Calcein-AM (1 mM DMSO solution).
9. Ethanol for disinfection.

2.5 Others

1. Silicone grease (Dow Corning, H.V.G).
2. Fine rod for scraping and spreading silicone grease (*see* **Note 8**).
3. Black membrane filter (Sartorius, 13006-025N).
4. Sprayer containing ethanol for disinfection.
5. Egg stand.
6. 35 mm Culture dish (Corning).
7. 50 mL Conical tube and rack.
8. Petri dish (120 mm in diameter).
9. Filter paper (Whatman, 55 mm in diameter, 1001-055).
10. Scale (stainless, minimum scale: 0.5 mm).
11. Razor blade (0.1 mm in thickness).
12. Blade holder.
13. Parafilm.
14. Aluminum block (25 mm in diameter, 15 mm in thickness).
15. Fine tungsten rod (electrolytic polished).
16. Humidified thermostat chamber (*see* **Note 9**).
17. Perfusion system by gravity on the stage of the upright microscope (*see* **Note 10**).

3 Methods

Preparation of retinal strips, culture, and observation of axons are completed within the dark clean room. Main procedures are described in the method section of [9].

3.1 Preparation of DMEM and Culture Chamber

1. Add DMEM (with phenol red) to 3.6 mL of serum mixture in a 50 mL conical tube so that the total volume becomes 30 mL (final serum concentrations: FBS 10%, chicken serum 2%).
2. Put the serum mixture in ice.
3. Put the conical tube in a dry incubator.
4. Coat the both walls of the trough of the culture chamber with silicone grease using a fine rod (*see* **Note 11**).
5. Secure a cover glass (25 mm in diameter) to either side of the culture chamber with silicone grease to make the bottom of the trough.
6. Put a small amount of silicone grease on the upper surface of the culture chamber along the both walls of the trough to secure another cover glass (15 mm in diameter).

3.2 Preparation of Matrigel® and Cooling

1. Put an aliquot of Matrigel® (152 μL) in a refrigerator at 4 °C.
2. After 30 min, put the aliquot of Matrigel® in ice.
3. Add 48 μL of serum mixture to the Matrigel® (final serum concentrations: FBS 20%, chicken serum 4%) (*see* **Note 12**).
4. Put the Matrigel® in ice.
5. Put an aluminum block (25 mm in diameter, 15 mm in thickness) in ice.
6. Put 35 mm culture dish on ice, and fill it with DMEM (with phenol red) (*see* **Note 13**).

3.3 Isolation of the Retina

1. Bring eggs incubated for 6 days to the clean room.
2. Put the egg on an egg stand with the air chamber up.
3. Break the shell, and expose the embryo in a petri dish.
4. Cut the neck, and transfer the head to a 35 mm dish filled with HBSS using a ring of stainless wire (*see* **Note 14**).
5. Enucleate an eye under a dissecting microscope, and transfer it by holding the lens with a fine forceps to another 35 mm dish filled with HBSS.
6. Remove extraocular muscles with a pair of fine forceps.
7. Make a small cut along the equator near the optic fissure.
8. Transfer the eye to a 35 mm dish filled with HBSS containing a black membrane filter.

9. Peel the pigment epithelium with a pair of fine forceps (*see* **Note 15**).
10. Cut the optic nerve head with a fine forceps.
11. Remove the optic nerve and the pigment epithelium.
12. Make a continuous cut along the equator to separate the anterior retina from the posterior retina.
13. Put the eye on the black membrane filter with the lens up.
14. Gently push the eye onto the membrane filter to attach the outer limiting membrane of the posterior pole to the membrane filter.
15. Separate the vitreous body from the posterior retina, and attach the posterior retina to the membrane filter by pressing the periphery on the membrane filter (*see* **Note 16**).
16. Attach the whole posterior retina to the membrane filter by pressing the periphery on the membrane filter.
17. Remove the vitreous body and the anterior retina together with the lens.

3.4 Preparation of Retinal Strips

1. Lift up the retina-membrane assembly from HBSS, and put it on a filter paper with the retinal side up.
2. Put the filter paper under the dissecting microscope to position the optic fissure vertically.
3. Put a scale near the retina on the left in parallel with the optic fissure (*see* **Note 17**).
4. Define the level of the optic nerve head by seeing the scale.
5. Make a horizontal cut at the level of the optic nerve head in the nasotemporal direction (i.e., perpendicularly to the optic fissure) with a piece of razor blade (*see* **Note 18**).
6. Make a horizontal cut 0.7–1.0 mm dorsal to the previous cut in parallel with it (*see* **Note 19**).
7. Make a horizontal cut 0.8–1.0 mm dorsal to the previous cut in parallel with it (*see* **Note 20**).
8. Make a vertical cut 1.2 mm left to the optic fissure.
9. Make an oblique cut 1.2–1.5 mm right to the optic fissure to make the ventral edge longer (*see* **Note 21**).
10. Make a drop of DMEM (25 μL) on Parafilm (*see* **Note 22**).
11. Transfer the retinal strip into the drop of DMEM (*see* **Note 23**).

3.5 Culture of Retinal Strip

1. Put the ice-cooled aluminum block in a petri dish, and fill the space around the aluminum block with ice.
2. Put the culture chamber on the aluminum block.

3. Drop 1 μL of ice-cooled Matrigel® (containing serums). Scrape the bottom of the trough with a fine tungsten rod to coat the bottom with the Matrigel®.
4. Put a wedge of membrane filter on the bottom of the trough (*see* **Note 24**).
5. Put the retinal strip on the bottom of the trough with the retina up.
6. Position the retinal strip perpendicularly to the wall of the trough using a fine tungsten rod (*see* **Note 25**).
7. Fix the retinal strip by the wedge of membrane filter.
8. Drop 8 μL of Matrigel® (containing serums) over the retinal strip (*see* **Note 26**).
9. If air bubbles are seen in the Matrigel®, remove them using a fine tungsten rod.
10. Put a cover glass (15 mm in diameter) over the retinal strip, and secure it to the upper surface of the culture chamber with silicone grease.
11. Put the culture chamber in a humidified thermostat chamber at 37 °C for gelling.
12. After 5 min, put the culture chamber on an acrylic disc (*see* **Note 27**).
13. Take out the DMEM containing 10% FBS and 2% chicken serum from the dry incubator.
14. Fill the trough of the culture chamber with the DMEM using a Pasteur pipette (*see* **Note 28**).
15. Put the acrylic disc with the culture chamber in a well of a six-well culture plate.
16. Fill the neighboring two wells with DMEM containing 10% FBS and 2% chicken serum.
17. Bridge the both ends of the trough to the neighboring wells with a pair of glass tubes (3 mm in diameter) containing the culture medium (*see* **Note 29**).
18. Fill the other wells with distilled water.
19. Inject 1 mL of distilled water into the well containing the chamber around the acrylic disc.
20. Put the six-well culture plate in an airtight jar.
21. Put CULTUREPAL® in the airtight jar.
22. Close the lid of the airtight jar.
23. Put the airtight jar in a dry incubator.
24. Incubate it for 24 h at 37.5–37.9 °C.

3.6 Observation of Growing Axons

1. Take out the airtight jar from the dry incubator.
2. Open the lid of the jar, and take off CULTUREPAL®.
3. Take out the six-well culture plate from the jar.
4. Remove the two glass tubes from the culture chamber.
5. Take out the culture chamber with the acrylic disc from the well, and put them on Kimwipe.
6. Separate the culture chamber from the acrylic disc.
7. Place the culture chamber on the stage of an upright fluorescence microscope.
8. Perfuse the trough of the chamber with HEPES-buffered DMEM without phenol red for 5 min at 2 mL/min.
9. Perfuse the trough of the chamber with HEPES-buffered DMEM without phenol red containing calcein-AM (10 μM, 10 mL) for 5 min at room temperature.
10. Wait for 30 min at room temperature.
11. Wash out the calcein-AM-containing solution for 5 min with HEPES-buffered DMEM without phenol red.
12. Fill the space between the cover glass over the retinal strip and the water immersion objective with HEPES-buffered DMEM without phenol red for high-magnification observation.
13. Take photos by fluorescence microscope (*see* **Note 30**).
14. To quantify the growth of RGC axons, measure the fluorescence intensity at a defined area by using ImageJ (*see* **Note 31**).

4 Notes

1. Put a dish filled with distilled water, to which SDS pellets (SERVA) are added to kill bacteria.
2. Wash the eggs with water. Put them on rack in the refrigerator so that the air chamber (round side of the egg) is up.
3. Purchase a kit. DIY by jointing and connecting tubes to make framework and covering it with a blackout curtain.
4. Make from acrylic disc. Refer to Supplementary Fig. S1a of [9].
5. Refer to Supplementary Fig. S1b of [9].
6. CULTUREPAL® produces CO_2 and maintains 5% CO_2 in the airtight jar.
7. Prepare aliquot of Matrigel® (152 μL) by slowly dispensing. Store at −20 °C before use.
8. Wooden toothpick. Clean it with Kimwipes and ethanol before use.

9. Put an aluminum block (19 mm in diameter, 22 mm in thickness) in the largest well (37 mm in diameter, 28 mm in depth) of Fibrinotherm (Baxter). Fill the space around the aluminum block with distilled water. Put a cover of plastic dish on the well to keep humidity.
10. Conventional perfusion system for electrophysiological recording.
11. Without this procedure, Matrigel® adheres to the wall of the trough.
12. If drug is applied, add it to the Matrigel®. Cool in ice, and mix them by tapping. Repeat cooling and tapping at least three times.
13. Cool a yellow tip by sucking the ice-cooled DMEM before sucking Matrigel®.
14. Make a ring (6–7 mm in diameter) of stainless wire by twisting. Connect it to a holder. Scoop a chicken head with the ring. Compare several heads for staging.
15. Insert the tip of forceps into the cleft between the retina and the pigment epithelium to separate them.
16. The vitreous body can be separated from the retina at the dorsal, nasal, and temporal parts. Then, separate the vitreous body from the optic fissure.
17. To measure lengths, *see* Fig. 1a.
18. By this cutting, the level of the optic nerve head is defined on the scale, *see* Fig. 1a.
19. This cutting makes the ventral edge of the retinal strip, *see* Fig. 1a.
20. This cutting makes the dorsal edge of the retinal strip and defines the width of the retinal strip, *see* Fig. 1a.
21. This oblique cut defines the direction of the retinal strip, *see* Fig. 1a.
22. Put a piece of Parafilm in 35 mm dish. If drug is applied, add the drug to the DMEM for preincubation.
23. Hold the lower right corner of the retinal strip with fine forceps.
24. This wedge is used to fix the retinal strip, *see* Fig. 1b.
25. Attach the membrane filter to the bottom to reduce the space between them. Unless, the retinal strip floats.
26. Suck 11 μL and leave 3 μL in the yellow tip not to add air bubbles.

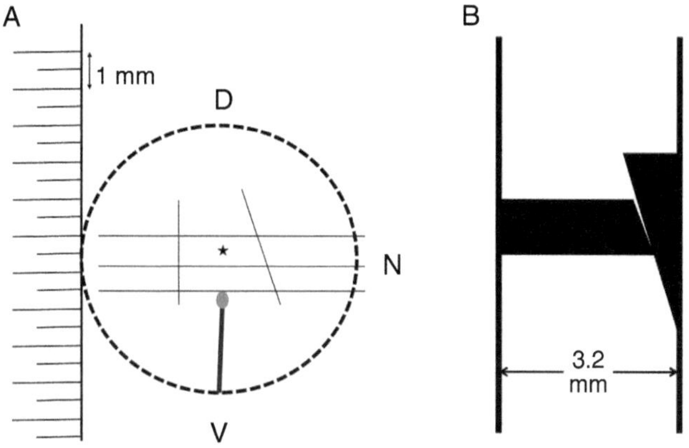

Fig. 1 Preparation of retinal strip. (**a**) Making cuts seeing a scale put on the left. *D* dorsal, *N* nasal, and *V* ventral. Dashed circle: posterior retina attached to membrane filter by pressing the periphery. Gray thick vertical line: optic fissure. Oval: optic nerve head. A star indicates the retinal strip used for study. (**b**) Arrangement of the retinal strip and a wedge of membrane filter on the bottom of the trough of culture chamber

27. Put a paper (25 mm in diameter) between the culture chamber and the acrylic disc to separate them. The papers between black membrane filters stored in a plastic case are used.
28. Carefully inject the DMEM from either end of the trough. Avoid air bubbles.
29. Refer to Supplementary Fig. S2a of [9].
30. It is absolutely necessary that the retinal strip is completely flat mounted on the membrane filter during incubation. When the retinal strip is detached from the membrane filter and curls up on it, axons extend in all directions (Fig. 2a). Folding of the retinal strip changes directions of outgrowing axons (Fig. 2b). When the retinal strip is detached from the membrane filter at the periphery, axons extend upward from the rounded edges (Fig. 2c, d).
31. For example, the fluorescence intensity between the lines 150 and 400 μm ventral to the retinal strip gives the amount of the axons that have extended ventrally from the retinal strip. Use the ratio of the fluorescence intensity against the background intensity measured at the wedge of membrane filter. Confocal fluorescence microscope and a low-magnification objective (×4) are used.

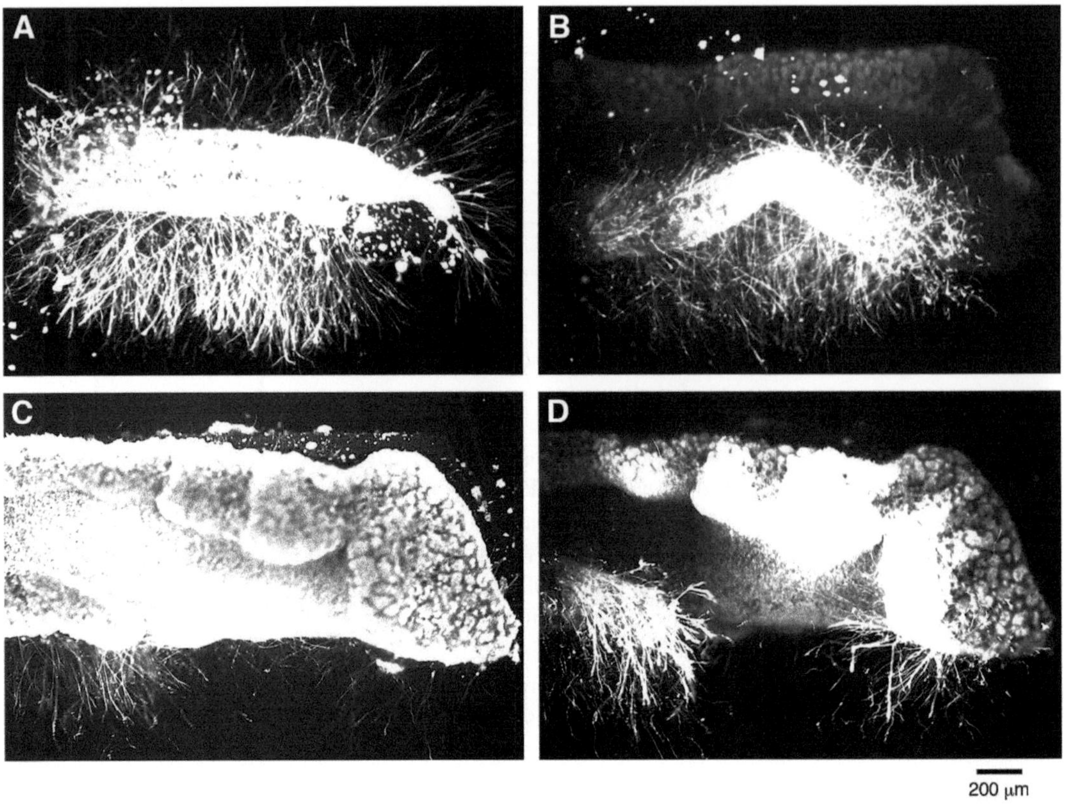

Fig. 2 Irregularly growing RGC axons. (**a**) Retinal strip was detached from membrane filter and curled up on it. (**b**) The central part of the ventral edge was detached from membrane filter and lifted up. (**c**, **d**) A retinal strip detached from membrane filter at the periphery. (**c**) The dorsal, nasal, and ventral edges were detached from membrane filter. (**d**) Axons extended from upper parts of the retinal strip

References

1. Halfter W, Dong S, Schurer B et al (2005) Embryonic synthesis of the inner limiting membrane and vitreous body. Invest Ophthalmol Vis Sci 46:2202–2209
2. Randlett O, Poggi L, Zolessi FR et al (2011) The oriented emergence of axons from retinal ganglion cells is directed by laminin contact in vivo. Neuron 70:266–280
3. Halfter W, Willem M, Mayer U (2005) Basement membrane-dependent survival of retinal ganglion cells. Invest Ophthalmol Vis Sci 46:1000–1009
4. Halfter W, Dong S, Balasubramani M et al (2001) Temporary disruption of the retinal basal lamina and its effect on retinal histogenesis. Dev Biol 238:79–96
5. de Curtis I, Reichardt LF (1993) Function and spatial distribution in developing chick retina of the laminin receptor $\alpha_6\beta_1$ and its isoforms. Development 118:377–388
6. Hynes RO (2002) Integrins: bidirectional, allosteric signaling machines. Cell 110:673–687
7. Srichai MB, Zent R (2010) Integrin structure and function. In: Zent R, Pozzi A (eds) Cell-extracellular matrix interactions in cancer. Springer, Berlin
8. Kim S-H, Turnbull J, Guimond S (2011) Extracellular matrix and cell signalling: the dynamic cooperation of integrin, proteoglycan and growth factor receptor. J Endocrinol 209:139–151
9. Yamashita M (2015) Weak electric fields serve as guidance cues that direct retinal ganglion cell axons in vitro. Biochem Biophys Rep 4:83–88

Chapter 6

Multiple Approaches for Enhancing Neural Activity to Promote Neurite Outgrowth of Retinal Explants

Chuan-Chin Chiao, Chin-I Lin, and Meng-Jung Lee

Abstract

Activity is important for neural development and regeneration. Enhancing neural activity can facilitate axon regrowth of retinal ganglion cells. Here, we describe various methods, including electrical stimulation, pharmacological manipulation, and optogenetics, to elevate neural activity of retinal explants in mice and to analyze their effects on promoting neurite outgrowth in organotypic culture.

Key words Retinal ganglion cells, Axon regeneration, Electrical stimulation, Optogenetics, Organotypic culture

1 Introduction

Neurons in the mammalian central nervous system (CNS) rarely regenerate and typically die soon after injury [1, 2]. However, the peripheral nervous system and developing mammalian CNS remain the ability of regeneration. It is known that neural activity plays an important role in early CNS development [3]. The correlated spontaneous activity during retinal development, also known as retinal waves, is thus thought to be responsible for the regenerative capacity of retinal ganglion cells (RGCs) [4, 5]. We hypothesize that exogenously driven neural activity that mimics endogenously occurring neural activity during retinal development, such as via electrical stimulation, pharmacologically removing inhibition, or light-activated channelrhodopsin-expressing RGCs, can enhance survival and neurite outgrowth of retinal explants. We have developed multiple approaches to demonstrate the effect of neural activity on facilitating axon growth of RGCs. Furthermore, ex vivo culture of mammalian neuroretina provides an attractive model for studying mechanisms of human retinal injury and degenerative disease [6–9]. Here, we describe both method of retinal explant culture and various means of enhancing neural activity.

Chai-An Mao (ed.), *Retinal Development: Methods and Protocols*, Methods in Molecular Biology, vol. 2092, https://doi.org/10.1007/978-1-0716-0175-4_6,

2 Materials

2.1 Animals

The experiments were performed on postnatal day 5 and 11 C57BL/6 mice and Tg (Thyl-COP4/EYFP)9Gfng (Thyl-ChR2) mice of either sex. The C57BL/6 mice were obtained from the National Laboratory Animal Center in Taiwan, and the ChR2 mice were originally obtained from the Jackson Laboratory in the USA.

2.2 Culture Medium

The contents of the culture medium (20 mL) are listed below:

1. Neurobasal-A
2. 0.6% Glucose
3. 1 mM Sodium pyruvate
4. 10 mM HEPES
5. 1× B-27
6. 2 mM L-Glutamine
7. 100 μg/mL Penicillin
8. 2.5 μg/mL Insulin
9. 6 mM Forskolin.
10. 100 nM IGF-1 or 50 ng/mL BDNF

3 Methods

3.1 Preparation of Retinal Explant

Retinas were isolated from postnatal *C57BL/6* mice or Tg (Thyl-COP4/EYFP)9Gfng (Thyl-ChR2) mice. The mice were deeply anesthetized and sacrificed by intraperitoneal injection of an overdose of 10 mg/kg ketamine and 10 mg/kg xylazine. To isolate the retinas, the eyeballs were enucleated using surgical scissors (Fig. 1) and bathed in 35 °C oxygenated (95% O_2 and 5% CO_2) Ames' medium (A1420; Sigma-Aldrich, St. Louise, MS, USA) containing 23 mM $NaHCO_3$. Using a dissection microscope, the eyeballs were hemisected around the ora serrata with fine scissors, and the lenses were removed immediately (Fig. 2). The retinas were then gently separated from the posterior eyecup, and the vitreous humor was removed carefully by Dumont forceps (Fig. 3). Finally, the isolated retinas were cut into four pieces (Fig. 4), and attached ganglion cell sides down onto Cell-Tak (354240; BD Biosciences, Franklin Lakes, New Jersey, U.S.)-coated coverslips for retinal explant culture (Fig. 5). All the procedures were performed under normal indoor illuminance.

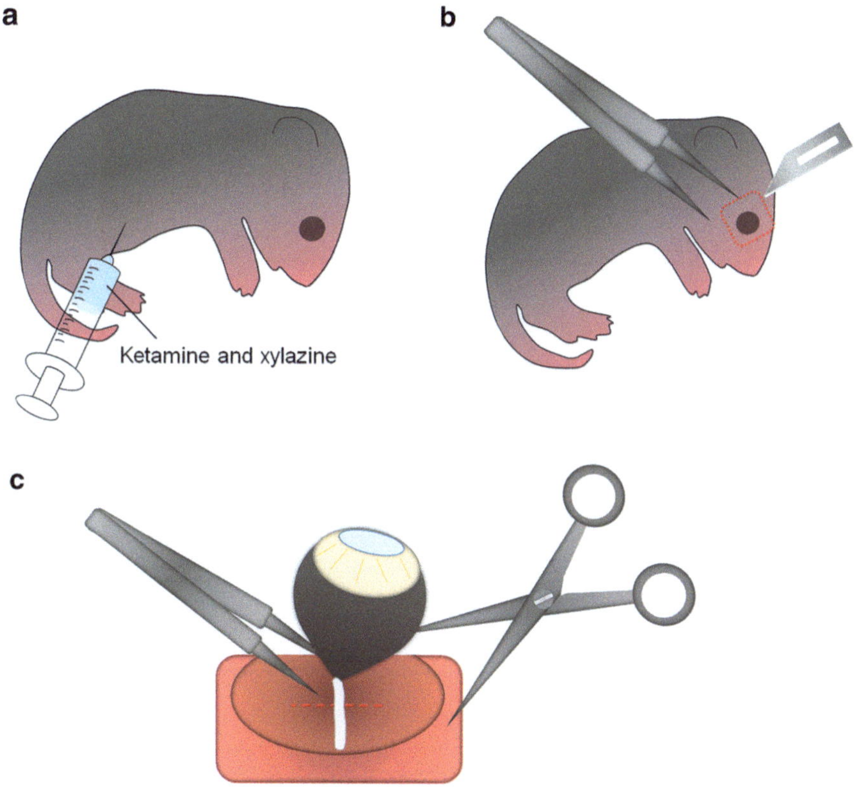

Fig. 1 Isolation of the eyeball from the postnatal mouse. (**a**) Anesthetize the postnatal mouse with overdose of ketamine and xylazine by intraperitoneal injection. (**b**) Cutting and stripping the eyelids with a blade. The red-dotted frame represents the eyelid to be isolated. (**c**) Pull out the eyeball with forceps, and then cut the optic nerve with surgical scissors. The red-dotted line represents the position of the cut

3.2 Preparation of Culture Medium

Preparation of the culture medium must be carried out in the laminar flow cabinet. The protocol is described below:

1. Prepare all the required ingredients for the culture medium into solutions at room temperature. The respective concentrations are at **step 3**. Vortex the reagent if necessary.
2. To prepare final volume of 20 mL of culture medium. First, put 0.12 g glucose into a 50 mL falcon tube with 18.6 mL Neurobasal-A.
3. Add 50× B27 (400 μL), 100 mM sodium pyruvate (200 μL), 1 M HEPES (200 μL), 200 mM L-glutamine (200 μL), 10 mg/mL penicillin (200 μL), 0.5 mg/mL insulin (5 μL), and 30 mM forskolin (4 μL) into Neurobasal-A.
4. Filter the mixed medium with a syringe through the filter of pore size 0.20 μm (Minisart® NML Syringe Filter 16534) into a new, sterilized falcon tube.
5. Add 100 μg/mL IGF-1 (156 μL) or BDNF after filtration.

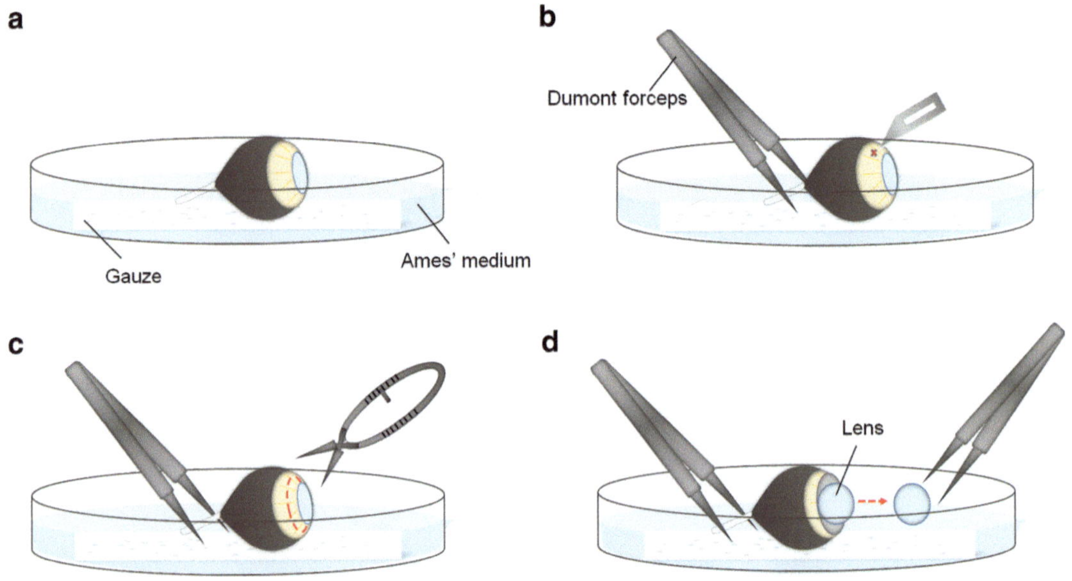

Fig. 2 Dissection of the eyeball. (**a**) Place the eyeball on the gauze in a dish. (**b**) Use forceps to hold the optic nerve and stabilize the eyeball, and then use a blade to create a cut on the cornea. The red mark represents the cut. (**c**) Use micro scissors to cut around the cornea. The red-dotted circle represents the cutting area. (**d**) Pull out the lens with Dumont forceps

6. Store the medium at 4 °C.
7. The medium has to be warmed up to room temperature (25 °C) before adding to the retinal explants.

3.3 Retinal Explant Culture

The retinal explants were placed in a 12-well plate and cultured in a 5% CO_2 humidified incubator at 35 °C for 5 days (Fig. 6a). All retinal explants were supplied daily with fresh culture medium (*see* the preparation above). Depending on the experiment, some retinal explants were either electrically stimulated using MEA for 1 h right before the culture or light stimulated using blue LED light from below for 1 h at the beginning of culture (Fig. 6). Alternatively, some retinal explants were activated with pharmacological treatment.

3.4 Immunohistochemistry

The cultured retinal explants were fixed with 4% paraformaldehyde and 0.1% glutaraldehyde for 1 h at room temperature and then rinsed three times in phosphate-buffered saline (PBS), 10 min each time. Next, the fixed explants were blocked by 4% normal donkey serum and 0.1% Triton X-100 in PBS for 1 h at room temperature to reduce nonspecific binding. The primary antibodies against class III beta-tubulin (TUJ1; 1:500; MMS-435P; Covance, Princeton, NJ, USA), MAP-2 (1:50; sc-20172; Santa Cruz, Finnell Street Dallas, TX, USA), and Tau (1:50; sc-1995, Santa Cruz) were then incubated with the explants overnight at 4 °C. After rinsing

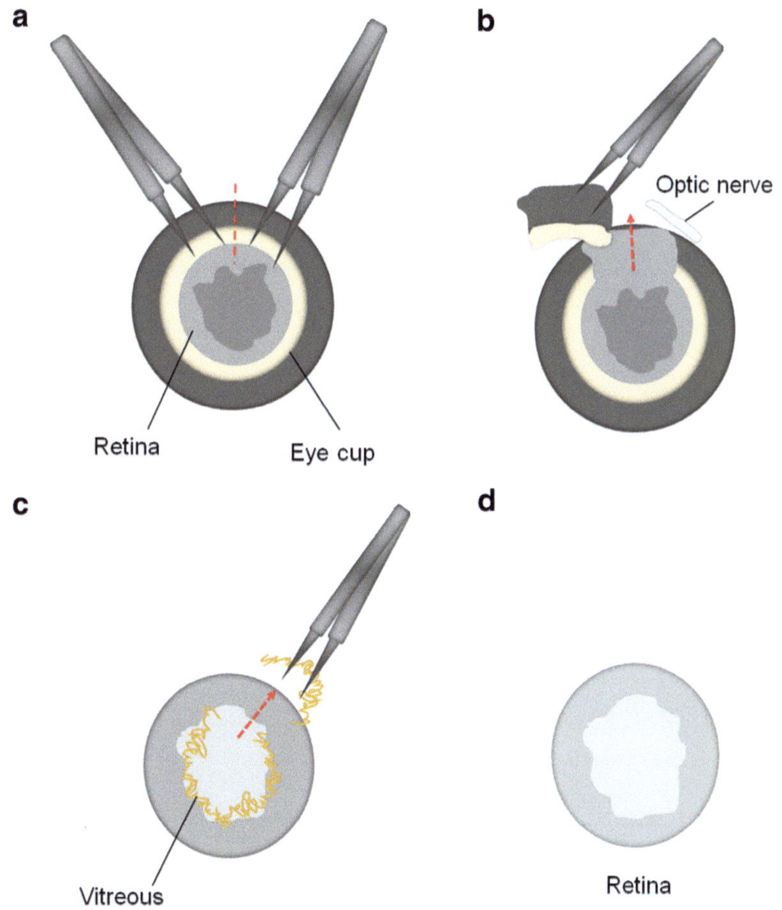

Fig. 3 Isolation of the retina. (**a**) Top view of the eyecup with the retina inside. Use one Dumont forceps to hold the eyecup and the other Dumont forceps to isolate the retina. The red-dotted line represents the starting point of peeling. (**b**) Carefully peel off the eyecup, and isolate the retina from the retinal pigment epithelium with two Dumont forceps, and then cut the optic nerve at the optic disk. (**c**) Use two Dumont forceps to carefully remove the vitreous humor. (**d**) A complete isolated retina

several times in PBS, the explants were incubated with the corresponding secondary antibody (1:250; Alexa 488, DyLight 549 or DyLight 649; Jackson Lab, Franklin, TN, USA) overnight at 4 °C. Finally, the explants were mounted in mounting medium with DAPI (Vector Laboratories, Burlingame, CA, USA; Fig. 7).

3.5 Neurite Outgrowth Quantification

Images of neurite outgrowth of the retinal explants were acquired using either an inverted microscope (Axio Observer.Z1, Zeiss, Germany) or a confocal microscope (LSM510, Zeiss). Based on the confocal images, we confirmed that beta III tubulin (TUJ1) was expressed in the neurites growing out from the RGCs but not from the glial cells in retinal explants (Fig. 8a, b). TUJ1 staining in retinal

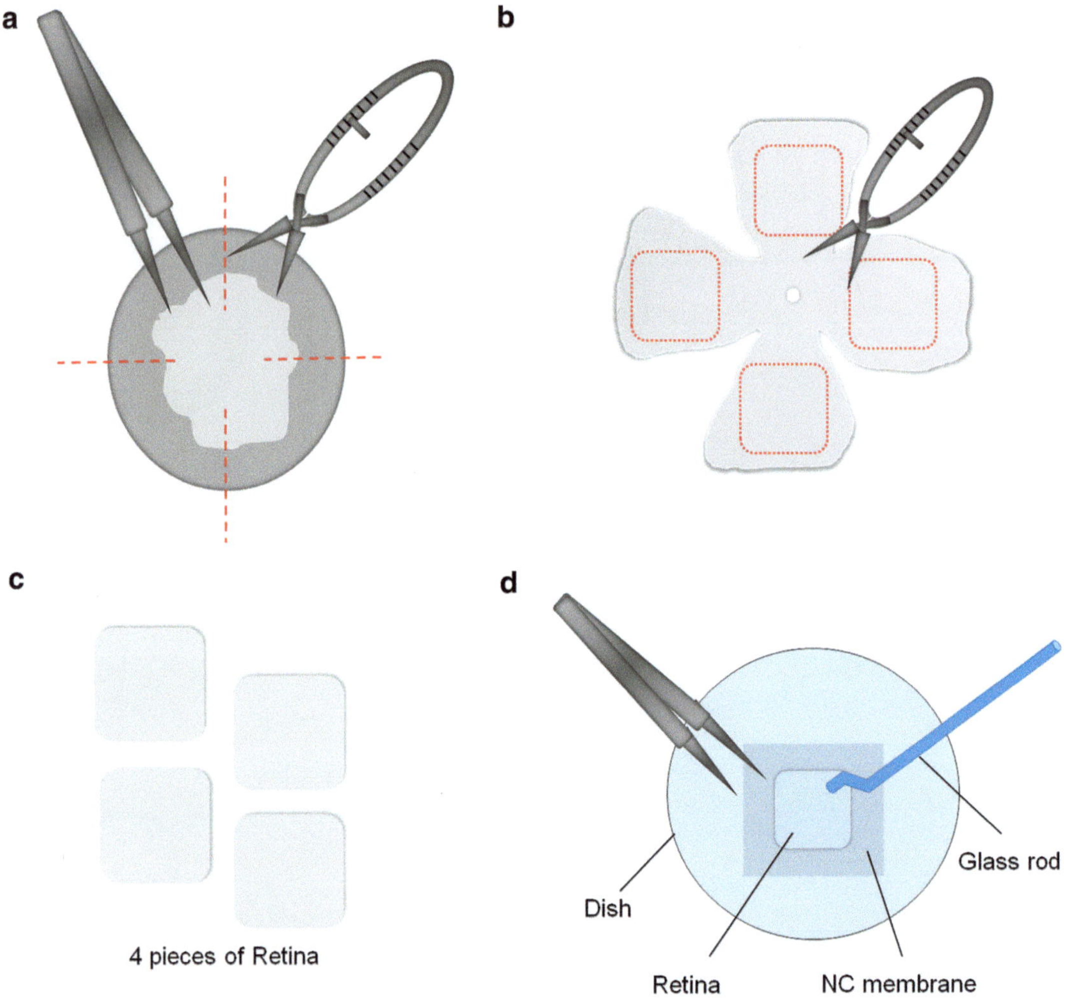

Fig. 4 Dividing and handling of the retina. (**a**) Use one Dumont forceps to fix the retina gently, and use micro scissors to make four cuts along the red-dotted line. (**b**) Use micro scissors to cut the retina into four small pieces. The red-dotted frames represent the cut retinal pieces. (**c**) Four trimmed pieces of the retina. (**d**) Use Dumont forceps and a glass rod to attach the retinal piece with the photoreceptor side down onto the NC membrane

flat mounts was shown to be RGC specific and was efficient in RGC identification [10–12]. Furthermore, although beta III tubulin is known to be expressed in both Tau- and MAP2-positive neurites, it was found that mostly processes longer than 200 μm were Tau positive, suggesting that they were primarily the grown axons of RGCs not the axons from amacrine cells (Fig. 8c). To quantify the extent of neurite outgrowth, images of the retinal explants from the inverted microscope were characterized and measured directly, while those images collected using the confocal microscope were analyzed via the following specific steps (Fig. 9). The confocal images were first split into different color channels (DAPI and

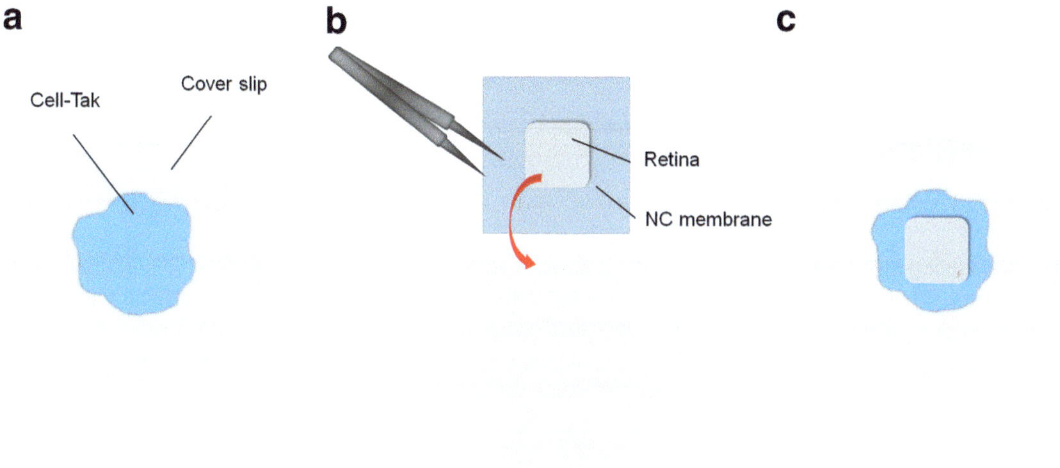

Fig. 5 Attaching the retina onto the coverslip. (**a**) Apply Cell-Tak to the coverslip, and wait a minute for air-dry. (**b**) Use forceps to pick up the NC membrane with the retinal piece, and attach the retinal explant with the ganglion cell side down onto the coverslip. (**c**) Remove the NC membrane, and the retinal explant is attached onto the coverslip

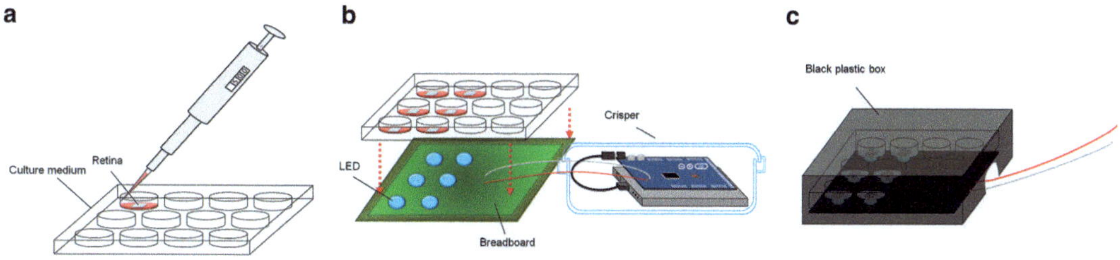

Fig. 6 Retinal explant culture and blue light stimulation. (**a**) Place the coverslip with the retinal explant into a 12-well plate, and slowly add the culture medium from the edge of the well. (**b**) Retinal explants are cultured in a 12-well plate and stimulated by the blue light LED array from below for only 1 h at the beginning of the experiment. The blue light LEDs (~680 cd/m^2, 470 nm) are powered and driven by an Arduino circuit in order to provide light stimulation, and the Arduino circuit with mobile power is placed in a container to keep dry in the incubator. (**c**) Cover the 12-well plate and the LED array with a black plastic box to prevent light leakage

TUJ1), which represent the area of the retinal explants and the area of the neurite outgrowth. Only neurites grown out from the explants were included in quantification, and these were defined as the total neurite area. The extent of neurite outgrowth was then characterized by dividing the total neurite area by the circumference of the explant. To further distinguish neurites of different lengths that are growing out from the explants, the neurite areas <100 μm, 100–200 μm, and >200 μm expanding outward from the contour of the explants were calculated separately. Note that the neurite area <100 μm can be considered as a measure of neurite density, whereas the neurite area >200 μm represents the amount

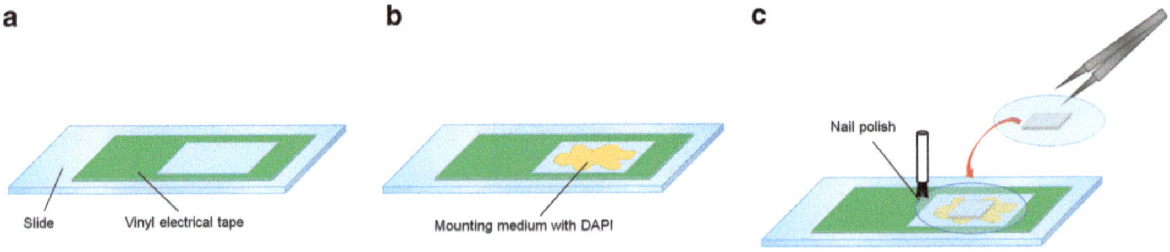

Fig. 7 Mounting the cultured retinal explant. (**a**) Attach the vinyl electrical tape onto the slide, and cut out a square area. (**b**) Add the mounting medium with DAPI to the square area. (**c**) Place the coverslip on the retinal explant located in the square area, and apply nail polish to seal the edge of the coverslip for confocal microscopy. The retinal explant has been immunostained for TUJ1 to visualize neurite outgrowth

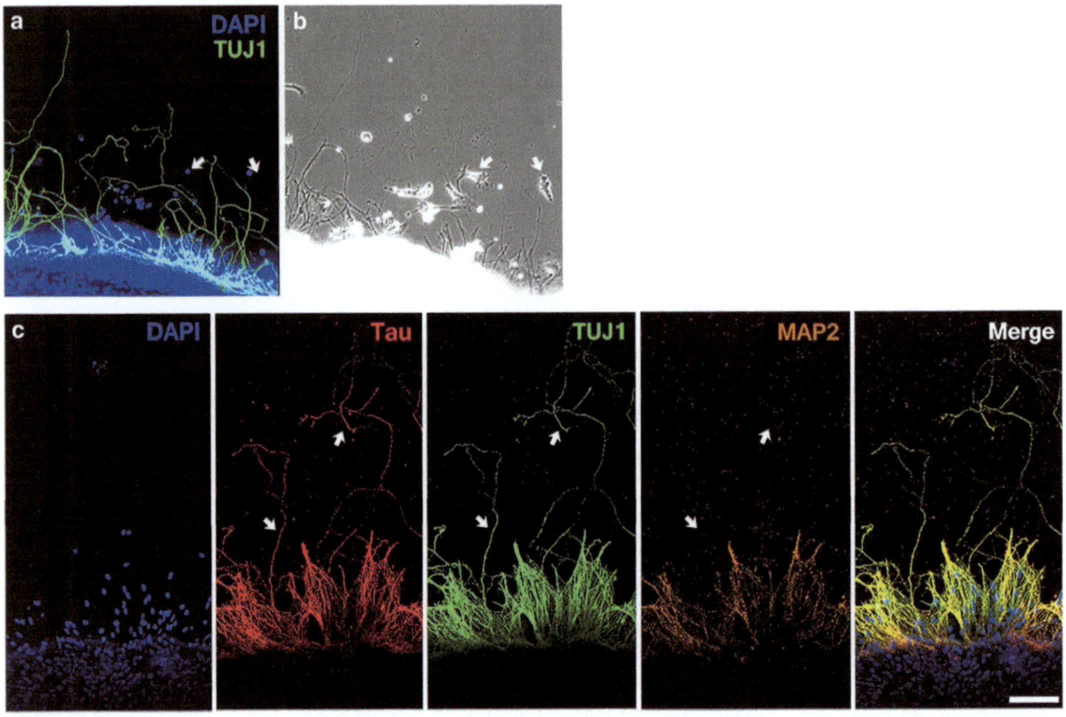

Fig. 8 Beta III tubulin (TUJ1) is expressed in neurites grown out from RGCs but not in glial cells from the retinal explants. (**a**) A confocal image and (**b**) a phase-contrast image of whole-mount retinal explants. All morphologically recognized neurites were TUJ1 positive (green), and the processes of glial cells (white arrows) were TUJ1 negative. DAPI was used to label nuclei (blue). (**c**) Beta III tubulin was expressed in both Tau- and MAP2-positive neurites, but neurites longer than 200 μm were Tau positive and MAP2 negative (white arrows). Scale bar, 100 μm

of elongated axons. A similar approach was used by Gaublomme et al. [13]. All image analyses were performed using ImageJ (National Institutes of Health, USA).

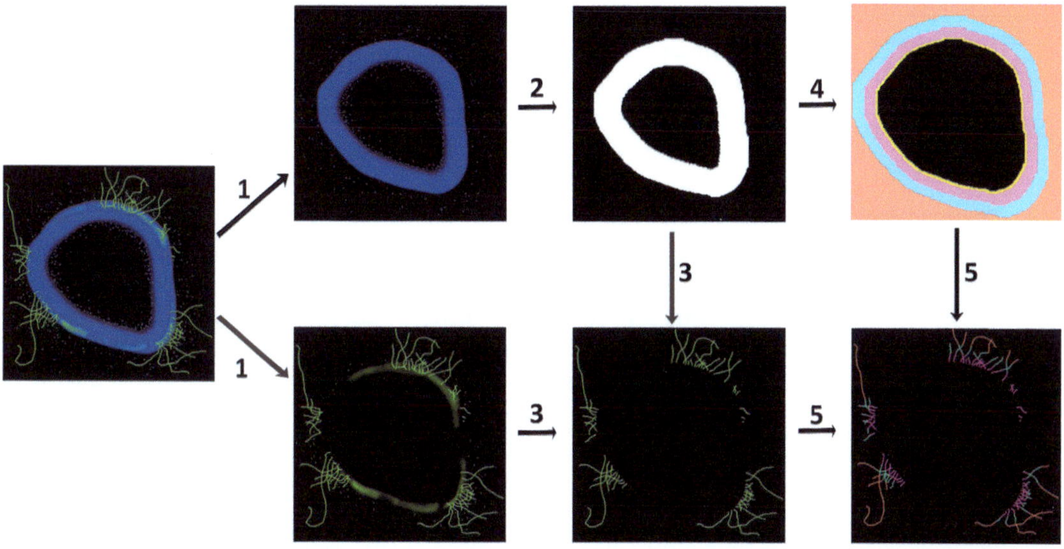

Fig. 9 Flow chart of the approach used to quantify neurite outgrowth of the retinal explant. The image of double immunostained retinal explant was (*1*) split into different color channels, green (TUJ1) and blue (DAPI). (*2*) The blue channel image was binarized, and the area of the retinal explant was defined. (*3*) Green (RGC neurites) signal inside the area of retinal explant was removed. (*4*) Two concentric outlines of an expanding circumference out from the retinal explant (100 μm interval) were delineated, and the area outside the explant was divided into three areas: 100 μm (pink), 100–200 μm (light blue), and >200 μm (orange). (*5*) Neurite areas (green channel) grown out from the explants were measured in these three areas separately

3.6 Cell Viability Assay

The condition of various retinal explants was examined using the Vybrant Apoptosis Assay Kit (Y3603; Thermo Fisher Scientific, Waltham, MA, USA). YO-PRO-1 in this assay is able to label both dead cells and cells undergoing apoptosis with green fluorescence [9]. The explants after being cultured for 5 days were incubated with 100 nM YO-PRO-1 for 1 h in a 37 °C and 5% CO_2 humidified incubator. The explants were then had images captured immediately using an inverted microscope (Axio Observer.Z1, Zeiss) at a fixed exposure time. The condition of each retinal explant was quantified by measuring the YO-PRO-1-positive area (fluorescence above the threshold) compared to the whole area of the explant. The specificity of RGC survival was examined by using a RGC-specific marker TUJ1 to establish a correlation between RGC survival and YO-PRO-1 signal. The results were then normalized against explants under the control condition. Image analysis was performed using ImageJ.

3.7 Electrical Stimulation

Retinal explants in the MEA were constantly perfused with fresh oxygenated Ames' medium (~1 mL/min) and kept at 31–33 °C. The explants were allowed to settle in the MEA for more than 30 min before electrical stimulation. Stimulation was performed using an in vitro USB-MEA-System (Multichannel Systems, Germany) with a MEA chip (60MEA200/10iR-ITO-pr-T) that consisted of 60 electrodes with diameters of 10 μm that were spaced 200 μm apart to form an 8×8 array. Different temporal patterns of pulsed stimuli (1 ms and 40 mV pulses) were applied to the explants for 1 h using 59 electrodes simultaneously with one electrode serving as ground via a stimulus generator (STG3000 Series).

3.8 Pharmacological Treatment

The retinal explants on coverslips were placed into a 6 mm culture dish with 5 mL of oxygenated and pre-warmed Ames's medium that had added to it some of the various pharmacological reagents, including picrotoxin (50 μM, Tocris), TPMPA (50 μM, Tocris), and strychnine (5 μM, Sigma) to remove GABA and glycine-mediated inhibition [5, 14, 15]. After 1 h of treatment, the explants were rinsed extensively with 30–50 mL of Ames's medium to remove any residual drugs. The concentrations of these pharmacological reagents were similar to the ones used in previous studies.

3.9 Optogenetics and Light Stimulation

A blue light LED array (~680 cd/m^2, 470 nm), which was powered and driven by an Arduino microcontroller board (Arduino MEGA 2560 rev3), was used to deliver light stimulation to the retinal explants from below (Fig. 6). There were three different temporal patterns of blue light used during the present study, namely, 5 Hz (100 ms pulse width), 20 Hz (25 ms pulse width), and 100 Hz (5 ms pulse width). By varying the pulse width according to the temporal frequency, the total number of photons delivered was kept the same for three different temporal patterns of light stimulation. To prevent the electronic circuit from being damaged in the incubator, the Arduino and the mobile power pack were put into a separate plastic container during the experiment (Fig. 6b). To avoid light leaking from the device, a light-sealed box was used to limit the light stimulation to that required for the experiments (Fig. 6c).

4 Notes

Part of these protocols has been published [16, 17]. Some of the results were presented in the Association for Research in Vision and Ophthalmology 2015 and 2018 annual meetings [18, 19].

Acknowledgments

We thank Dr. Chih-Tien Wang for kindly providing us with the recipe for culturing retinal explants. We also thank Dr. Ron Meyer and Ms. Jill Miotke for providing us with their laboratory protocols for the fixation and immunostaining of retinal explants. We appreciate Mr. Yueh-Chun Tsai in assisting us to assemble the Arduino microcontroller board for controlling blue light stimulation. This project was supported by the Ministry of Science and Technology in Taiwan, MOST-107-2311-B-007-002-MY3 (to C.C.C.).

References

1. Nicholls J, Saunders N (1996) Regeneration of immature mammalian spinal cord after injury. Trends Neurosci 19(6):229–234
2. Liu K, Tedeschi A, Park KK, He Z (2011) Neuronal intrinsic mechanisms of axon regeneration. Annu Rev Neurosci 34:131–152. https://doi.org/10.1146/annurev-neuro-061010-113723
3. Spitzer NC (2006) Electrical activity in early neuronal development. Nature 444 (7120):707–712. https://doi.org/10.1038/nature05300
4. Firth SI, Wang CT, Feller MB (2005) Retinal waves: mechanisms and function in visual system development. Cell Calcium 37 (5):425–432. https://doi.org/10.1016/j.ceca.2005.01.010
5. Syed MM, Lee S, Zheng J, Zhou ZJ (2004) Stage-dependent dynamics and modulation of spontaneous waves in the developing rabbit retina. J Physiol 560(Pt 2):533–549. https://doi.org/10.1113/jphysiol.2004.066597
6. Li Y, Zhang Y, Qi SN, Su GF (2018) Retinal organotypic culture—A candidate for research on retinas. Tissue Cell 51:1–7
7. Rettinger CL, Wang HC (2018) Quantitative assessment of retina explant viability in a porcine ex vivo neuroretina model. J Ocul Pharmacol Ther 34(7):521–530
8. Koizumi A, Zeck G, Ben YX, Masland RH, Jakobs TC (2007) Organotypic culture of physiologically functional adult mammalian retinas. PLoS One 2(2):e221
9. Moritoh S, Tanaka KF, Jouhou H, Ikenaka K, Koizumi A (2010) Organotypic tissue culture of adult rodent retina followed by particle-mediated acute gene transfer in vitro. PLoS One 5(9):e12917
10. Yin YQ, Cui Q, Li YM, Irwin N, Fischer D, Harvey AR, Benowitz LI (2003) Macrophage-derived factors stimulate optic nerve regeneration. J Neurosci 23(6):2284–2293
11. Cui Q, Yip HK, Zhao RCH, So KF, Harvey AR (2003) Intraocular elevation of cyclic AMP potentiates ciliary neurotrophic factor-induced regeneration of adult rat retinal ganglion cell axons. Mol Cell Neurosci 22(1):49–61. https://doi.org/10.1016/S1044-7431(02)00037-4
12. Snow RL, Robson JA (1994) Ganglion-cell neurogenesis, migration and early differentiation in the chick retina. Neuroscience 58 (2):399–409. https://doi.org/10.1016/0306-4522(94)90046-9
13. Gaublomme D, Buyens T, Moons L (2013) Automated analysis of neurite outgrowth in mouse retinal explants. J Biomol Screen 18 (5):534–543. https://doi.org/10.1177/1087057112471989
14. Gavrikov KE, Dmitriev AV, Keyser KT, Mangel SC (2003) Cation-chloride cotransporters mediate neural computation in the retina. Proc Natl Acad Sci U S A 100 (26):16047–16052. https://doi.org/10.1073/pnas.2637041100
15. Farajian R, Pan F, Akopian A, Volgyi B, Bloomfield SA (2011) Masked excitatory crosstalk between the ON and OFF visual pathways in the mammalian retina. J Physiol 589 (18):4473–4489. https://doi.org/10.1113/jphysiol.2011.213371
16. Lee MJ, Chiao CC (2016) Short-term alteration of developmental neural activity enhances neurite outgrowth of retinal explants. Invest Ophthalmol Vis Sci 57(15):6496–6506. https://doi.org/10.1167/iovs.16-19854
17. Lin CI, Chiao CC (2019) Blue light promotes neurite outgrowth of retinal explants in postnatal ChR2 mice. eNeuro 6(4):0391–18.2019. https://doi.org/10.1523/ENEURO.0391-18.2019
18. Lee MJ, Chiao CC (2015) Electrical activity but not correlated neural activity promotes neurite outgrowth of mouse retinal explants. Invest Ophthalmol Vis Sci 56(7):4970
19. Lin CI, Chiao CC (2018) Blue light promotes neurite outgrowth of retinal explants in postnatal ChR2 mice. Invest Ophthalmol Vis Sci 59 (9):5500

Chapter 7

Adeno-Associated Virus as Gene Delivery Vehicle into the Retina

Shuyun Deng and Kazuhiro Oka

Abstract

Initially discovered as a contaminant of adenovirus preparations, adeno-associated virus (AAV) has proved one of the most promising viral vectors for human gene therapy. The safety profile of AAV has been well-characterized in vivo studies, and the first gene therapy for patients with vision loss caused by Leber congenital amaurosis or retinitis pigmentosa was approved by the US Food and Drug Administration in 2017. This is an exciting era for investigators working on retina biology and treatments for blindness. In this chapter, we provide detailed methods for laboratory-scale production, purification, and characterization of AAV.

Key words Adeno-associated virus, Serotype, Packaging, Purification, Titration

1 Introduction

The US Food and Drug Administration (FDA)'s approval of adeno-associated virus (AAV) gene therapy for retinal dystrophy associated with retinal pigment epithelial (RPE) 65 mutations is a landmark for therapeutic gene therapy. Despite controversies related to affordability, this approval argues for genetic manipulation as therapeutic options to treat eye diseases. AAV is a nonpathogenic single-stranded DNA virus, which was discovered as a contaminant in adenovirus [1]. AAV is a member of the Parvoviridae family with an icosahedral symmetric capsid approximately 20 nm in diameter. More than 100 unique AAV capsid sequences have been identified, and several AAV receptors and attachment factors have been identified (reviewed in [2]). The presence of these receptors and attachment factors in target cells determines cell-type- or tissue-specific transduction and therefore must be carefully considered. AAV2, AAV5, and AAV8 have been used for gene delivery to the retina. However, the *cap* gene encoding three capsid viral proteins (VP1, VP2, and VP3) is only 2.2 kb. Bioengineering of capsid proteins is relatively easy and thus an active research area

Chai-An Mao (ed.), *Retinal Development: Methods and Protocols*, Methods in Molecular Biology, vol. 2092,
https://doi.org/10.1007/978-1-0716-0175-4_7,

[3, 4]. Researchers should be vigilant in recent development for efficient gene delivery to target cells (reviewed in [5, 6]).

AAV is a leading candidate for therapeutic gene therapy owing to its safety profile and efficacy compared with other viral vectors [7]. AAV genome mostly exists as an extrachromosomal episome [8]; however, vector integration occurs at low levels, which may cause genotoxicity [9–11]. A significant limitation of AAV is the packaging capacity (<4.7 kb). Fast-onset self-complementally AAV (AAVsc) has even smaller capacity, only ~2.4 kb. The oversized AAV (>4.7 kb) can be packaged in heterogeneous size truncated at around 5 kb with poor production yield compared to regular AAV [12]. To overcome the size limitation, a gene is split in multiple AAVs. In a dual AAV system, the reconstitution of the full-length expression cassette is achieved after coinfection of the same cell by homologous recombination between overlapping regions contained in both the 5′ and 3′ genome [13]. Triple AAV vectors with a capacity of up to 14 kb have been reported, which delivered to the pig retina at efficiency up to 40% [14].

Our basic protocol provides instructions for generating recombinant AAV in standard molecular biology laboratories. AAV genomes can be packaged in HEK293 or HEK293T cells by triple plasmid transfection using calcium phosphate or polyethylenimine (PEI) transfection methods. We describe these two methods as well as other alternative transfection reagents in Subheading 4.2. AAV is present in both the medium and the cell lysate after transfection [15], which depends on serotypes [16]. Both fractions are collected, processed, and combined before purification. Purification of AAV is based on density of viral particles. In Subheading 4.4, we describe two methods, iodixanol and CsCl purification, which can be performed with standard laboratory equipment. AAV purified by these methods has sufficient purity for experiment. The physical titer of AAV is determined by QPCR. Infectious titer is influenced by the presence or absence of AAV receptors and attachments sites. In Subheading 4.5, the general method to determine infectious titer is provided. The principle can be applied for various AAV serotypes and target cells.

2 Materials

2.1 Molecular Biology

1. AAV transfer plasmid vector: many AAV plasmid vectors containing AAV inverted terminal repeats (ITR) are available through Addgene (https://www.addgene.org/).
2. AAV Rep-Cap plasmid: 1 through 6, DJ and DJ/8 can be purchased through Cell Biolabs, Inc. Other native serotypes such as 8 and 9 are available through University of Pennsylvania Vector Core (https://gtp.med.upenn.edu/labs/vector-core).

3. Helper plasmid containing adenoviruses E2A, E4, and VA can be purchased from the same source.
4. Subcloning efficiency competent *E. coli* DH5α cells or other cells not prone to recombination with transformation efficiency $>1 \times 10^6$ transformants/μg DNA/50 μL competent cells.
5. Appropriate restriction enzymes and ligase kit, or cloning kit.
6. LB/ampicillin liquid medium: LB medium containing 50 μg/mL ampicillin.
7. LB agar plate containing 100 μg/mL ampicillin.
8. Plasmid mini prep kit (available from many major molecular biology suppliers such as Omega Bio-tek, Qiagen, and Takara Bio USA).
9. Plasmid endotoxin-free midi or maxi kit (available from many major molecular biology suppliers) 1% agarose gel.
10. 3 M Sodium acetate and pH 5.2 (liquid is available from many major molecular biology suppliers such as Teknova).
11. 70% and 100% Ethanol.
12. Transfection-grade H_2O.
13. Agarose gel electrophoresis system.
14. Bacteria incubator.
15. Orbital shaker incubator for bacteria.

2.2 Packaging Cells

1. HEK293T cells: Derivative of human embryonic kidney (HEK) 293 cells which contain the SV40 T antigen. HEK293T cells can be purchased from ATCC (ATCC CRL-3216) and other manufactures such as Cell Biolabs and Takara Bio USA. For AAV production, HEK293T cell is popular.
2. HEK293 cells (alternative to HEK293T) can be purchased from ATCC (ATCC CRL-1573).
3. Complete growth medium: DMEM high glucose supplemented with 10% heat-inactivated FBS, pyruvate, glutamine, and 1× antibiotic–antimycotic (or antibiotic).
4. 0.25% Trypsin EDTA.
5. DMEM high glucose supplemented with 5% heat-inactivated FBS, pyruvate, glutamine, and 1× antibiotic–antimycotic.
6. 15 mL and 50 mL Conical tubes.
7. 15 cm Cell culture dish.

2.3 Transfection

1. 2× HEPES-buffered saline (2×HBS): Dissolve 5 g of HEPES and 8.475 g NaCl in 470 mL of H_2O, and adjust pH to 7.1 with 1 M NaOH, and then adjust the volume to 500 mL. Sterilize by filtration through a 0.22 μm filter. The pH may

be increased by 0.1–0.3 by filtration. Prepare 3–4 buffer with pH higher or lower to 7.1 (7.05–7.15), test by transfection, and pick the best one. Dissolve 5.01 g $Na_2HPO_4{\cdot}12H_2O$ (MW = 368.14), and make up to 200 mL, and sterilize by 0.22 μm filter. Mix 500 mL of 2× HEPES and 10 mL of 70 mM Na_2HPO_4. This is 2× HBS. Store at room temperature (good for at least 6 months).

2. 2.5 M $CaCl_2$ for calcium phosphate transfection: Dissolve 73.5 g $CaCl_2{\cdot}2H_2O$, and make up to 200 mL. Sterilize by 0.22 μm filter, and store at −20 °C. This solution should remain liquid at −20 °C.
3. Polyethylenimine (PEI) linear MW 25,000 (Polysciences, cat no. 23966): Dissolve 50 mg of PEI in 45 mL of PBS, adjust to a pH 4.5 with 0.3 N HCl, and bring the volume to 50 mL. Filter-sterilize using a 0.22 μm filter. Store up to 6 months at 4 °C, or store at −20 °C.
4. Opti-MEM.
5. iMFectin polyDNA transfection reagent (GenDEPOT cat# I7200, available through Thermo Fisher Scientific).

2.4 AAV Purification by Iodixanol

1. Cell suspension buffer (CSB): 50 mM Tris pH 8.0, 5 mM $MgCl_2$, 150 mM and NaCl.
2. CSB2 (CCB for AAV2): 50 mM Tris pH 8.5, 2.5 mM $MgCl_2$, and 500 mM NaCl (*see* **Note 1**).
3. 5% Sodium deoxycholate.
4. 2 M $MgCl_2$: Dissolve 38.084 g $MgCl_2$ in 200 mL of H_2O, and sterilize by autoclave or filtration.
5. 10 mg/mL DNase I in 20 mM Tris–HCl, 50 mM NaCl, 0.1 mg/mL BSA, 50% glycerol, and pH 7.5. Store at −20 °C.
6. 100 U/mL Benzonase (MilliporeSigma).
7. 10 mg/mL RNase A in 10 mM Tris–HCl, 15 mM NaCl, and pH 7.5. Store at −20 °C.
8. 40% Polyethylene glycol (PEG) 8000 and 2.5 M NaCl: Add 200 g PEG8000 to 250 mL of 5 M NaCl with cell culture-grade water, and add water to less than 500 mL. Leave at 37 °C for several hours to overnight while stirring on magnetic stirrer. Alternatively, PEG solution can be heated to 65 °C until PEG is dissolved. Bring the volume to 500 mL, and filter-sterilize. Pre-wet the filter with a small volume of 2.5 M NaCl prior to adding the PEG/NaCl solution (*see* **Note 2**).
9. HBS (AAV suspension solution): 50 mM HEPES, 150 mM NaCl, 1% sarcosyl, 20 mM EDTA, and pH 8.0.
10. OptiPrep: 60% iodixanol in water (MilliporeSigma, Stem Cell Technologies).

11. PBS-MK: 1× PBS, 1 mM $MgCl_2$, and 2.5 mM KCl.
12. Phenol red: 0.5% stock in PBS-MK.
13. Dulbecco's PBS (DPBS), no magnesium, and no calcium.
14. Cell scraper.
15. 250 mL Centrifuge tube.
16. Low-speed centrifuge with adaptor for 250 mL tube.
17. High-speed centrifuge.
18. Ultracentrifuge.
19. NVT65 or SW32 Ti rotor.
20. OptiSeal ultra-clear centrifuge tube for NVT65 (Beckman #362181).
21. Centrifuge tube for SW32 Ti (Beckman #326823).
22. 15 mL Centrifugal filter unit with Ultracel-100 membrane (Amicon Ultra-15 #UFC910024, 100,000 cutoff, 0.0062 μm).

2.5 AAV Purification by CsCl

1. 10× Tris density buffer (TD): Dissolve 80.0 g NaCl, 3.8 g KCl, Tris 30.0 g, and $Na_2HPO_4{\cdot}12H_2O$ 2.5 g (or $Na_2HPO_4{\cdot}7H_2O$ 1.87 g), adjust pH 7.4–7.5, and then make up to total volume of 1000 mL.
2. *d* = 1.3 g/mL CsCl solution in 1× TD: pour 350 mL of 1× TD into a beaker. Add 150 g CsCl. Mix with a stir bar, and equilibrate to room temperature. Place exactly 1 mL of the CsCl solution in a small weighing boat using a 1 mL pipet calibrated such that 1 mL of water is exactly equal to 1.00 g, obtain an accurate weight, and return the 1 mL aliquots to the beaker. If the density is higher than desired, add more TD to the beaker. If the density is lower, add more CsCl powder. Repeat this process until the exact density concentration is reached. Sterilize the density gradient solution using a 0.22 μm filter. The solution can be stored up to 6 months at room temperature.
3. *d* = 1.5 g/mL CsCl solution in 1× TD: Add 180 g to 350 mL of 1× TD, and follow the instruction above.

   Slide-A-Lyzer dialysis cassette (0.5–3 mL capacity, 10,000 Mw #66380).

2.6 Reagents for Titration

1. 2× SYBR Green PCR Master Mix
2. 10 pmol/μL qPCR primers (forward and reverse)
3. Ninety-six-well reaction plate
4. Caps for reaction plate
5. Real-time QPCR platform

3 Biosafety

According to the National Institutes of Health biosafety guidelines, AAV handling should be performed in a laboratory operating at Biosafety Level 2 (BL-2) approved by the user's Institutional Biosafety Committee.

The requirements include the use of laminar flow hoods, the establishment of proper procedures for decontamination and disposal of liquid and solid waste, and methods for disinfection of contaminated surface and equipment. All precautions for cell culture are applied. All solutions and equipment coming into contact with cells must be sterile, and proper aseptic technique should be used accordingly.

4 Methods

4.1 Clone Gene of Interest into AAV Transfer Vector

Molecular cloning of gene of interest into AAV transfer vector can be done by standard cloning methods. In-Fusion Cloning Kit from Takara (Clontech #639636) and GenBuilder DNA Assembly from GenScript (#L00701-10 or L00744-10) work well (*see* **Note 3**). After verifying by either enzyme digestion or sequencing analysis, DNA is prepared by an endo-free plasmid preparation kit. AAV inverted terminal repeat (ITR) cannot be sequenced accurately by Sanger sequence method; however, the presence of intact ITR can be indirectly confirmed by SmaI digestion. The following method is used for sterilization of DNA for transfection:

1. Add 1/10 volume of 3 M sodium acetate buffer (pH 5.2) to DNA solution (e.g., add 0.1 mL of sodium acetate buffer to 1 mL of DNA solution).
2. Add two volumes (2.2 volumes of original DNA volume) of 100% ethanol to the above solution (e.g., add 2.2 mL of 100% ethanol to 1.1 mL of DNA–sodium acetate solution).
3. Mix by inverting the tube several times.
4. Incubate at −20 °C for 1 h or until use.
5. Centrifuge the tube for 5 min at 10,000 × *g*, and remove the supernatant.
6. Rinse the DNA pellet with 5 mL of 70% ethanol.
7. Centrifuge the tube for 3 min at 10,000 × *g*; then move the tube under the tissue culture hood. DNA should be kept sterile from this step forward.
8. Remove the supernatant, briefly air-dry DNA, and resuspend in transfection-grade H_2O.
9. Measure DNA concentration by OD_{260nm}.

4.2 Transfect and Package AAV

4.2.1 Transfection

1. Aspirate medium, rinse with 10 mL of PBS, and overlay 1 mL of 0.25% trypsin EDTA. Cells should be 70–80% confluent on the day of transfection (*see* **Note 4**).
2. Incubate cells in CO_2 incubator for 1–2 min, and detach cells by tapping the dish.
3. Add 10 mL of complete growth medium for 1:5.5 split.
4. Mix cells by pipetting up and down several times to dislodge cells and plate 2 mL of cell suspension/15 cm dish.
5. Add 18 mL of complete growth medium to total 20 mL/ 15 cm dish.
6. Incubate cells at 37 °C in a humidified CO_2 incubator.

   Cells should be 70–80% confluence at transfection the next day.

4.2.2 Calcium Phosphate Precipitation

1. Prepare DNA solution for 10 × 15 cm dish in a 50 mL tube (50 μg of helper plasmid, 25 μg of Rep/Cap plasmid, 25 μg of AAV transfer plasmid, and 150 μL of 2.5 M $CaCl_2$ in total volume of 1.5 mL/dish).
2. Pipette 1.5 mL/dish of 2× HBS in a 50 mL tube (make for 10 × 15 cm dish).
3. Slowly add DNA solution to 2× HBS while vortexing the tube at slow speed.
4. Incubate at room temperature for 10–15 min.
5. Add 3 mL of DNA mix to each dish. Gently mix the plate up–down and right–left to evenly distribute the DNA mix over the entire plate. Do not swirl.
6. Incubate at 37 °C for 6 h in a humidified CO_2 incubator.
7. Remove DNA mix and medium by aspiration.
8. Add 20 mL of fresh DMEM/5% FBS/dish.
9. Incubate at 37 °C for 72 h in a humidified CO_2 incubator.

4.2.3 PEI Transfection

1. Prepare DNA solution (10 μg of helper plasmid, 5 μg of Rep/-Cap plasmid, 5 μg of AAV transfer plasmid in 0.5 mL of Opti-MEM/15 cm dish). Mix by vortex.
2. Add 80 μL of PEI solution/15 cm dish and mix by vortex.
3. Incubate for 10 min at room temperature.
4. Add mixture dropwise to each dish.
5. Incubate at 37 °C for 5–6 h in a humidified CO_2 incubator.
6. Aspirate medium, and add fresh 20 mL of DMEM/5% FBS/1× antimycotic–antibiotic.
7. Incubate at 37 °C for 72 h in a humidified CO_2 incubator.

4.2.4 iMFectin Poly

1. Aspirate medium, and feed with 9 mL of DMEM/5% FBS/1× antimycotic–antibiotic 30–60 min before transfection.
2. Prepare DNA solution (4 μg of helper plasmid, 2 μg of Rep/-Cap plasmid, 2 μg of AAV transfer plasmid in 0.5 mL of plain DMEM/dish).
3. In a separate tube, dilute 24 μL of iMFectin poly in 0.5 mL of plain DMEM/dish, and mix by vortex for 10 s.
4. Add diluted iMFectin reagent to DNA solution, mix by vortex for 10 s, and incubate for 15 min at room temperature (no longer than 30 min).
5. Overlay 1 mL of DNA/iMFectin mixture/dish.
6. Incubate at 37 °C for 4 h in a humidified CO_2 incubator.
7. Add 10 mL of transfection medium to each dish.
8. Incubate at 37 °C for 72 h in a humidified CO_2 incubator.

4.3 Harvest of Cells and Medium

During AAV production, AAV is secreted into culture medium. The proportion of secreted and cell-associated AAV depends on serotypes [16]. Cell-associated and secreted AAVs are collected separately and combined before purification.

4.3.1 Cell-Associated AAV

1. Check transfection efficiency under fluorescence microscope if AAV transfer vector contains fluorescence marker gene (*see* **Note 5**).
2. Use a cell scraper to scrape off the cells from each plate.
3. Transfer the cell suspension to a 250 mL centrifuge tube.
4. Centrifuge at 228 × *g* for 10 min at 20 °C.
5. Transfer the supernatant to a new 250 mL centrifuge tube (go to Subheading 4.3.2).
6. Loosen cell pellets by tapping the bottom of the centrifuge tubes against the metal surface of cell culture hood, resuspend cells in CCB (1 mL/plate), and transfer to a new tube (*see* **Note 6**).
7. Add 5% sodium deoxycholate (0.1 mL/mL of cell suspension) to cell suspension, and invert the tube gently.
8. Incubate at room temperature for 30 min on rocking platform, or mix periodically.
9. Add 2 M $MgCl_2$ (10 μL/mL of cell suspension).
10. Add DNase I (10 μL/15 cm dish transfected) and RNase A (10 μL/15 cm dish transfected).
11. Incubate at 37 °C for 1 h, and mix periodically.
12. Centrifuge the mixture at 5000 × *g* for 10 min at 4 °C.
13. Transfer the supernatant to a new tube (*see* **Note 7**).

4.3.2 AAV Secreted into Medium

1. Digest supernatant with DNase I and RNase A (10 μL/plate) at 37 °C for 1 h.
2. Add 40% PEG solution to the supernatant to final concentration of 8% (e.g., 50 mL of 40% PEG for every 200 mL supernatant).
3. Mix well by inverting tubes several times, and leave it in a refrigerator overnight or until purification (*see* **Note 8**).
4. Centrifuge at 2800 × *g* for 30 min at 4 °C.
5. Aspirate the supernatant.
6. Loosen the pellets by hitting the bottom of the centrifuge tube against the metal surface of cell culture hood.
7. Resuspend PEG pellets in HBS (3 mL for the pellets collected from 200 mL medium corresponding to 10 × 15 cm dish). Mix on the rocker platform for 1 h or until the pellet is completely dissolved (*see* **Note 9**).

4.4 Purification

Discontinuous iodixanol or CsCl density gradient ultracentrifugation are standard methods for purification. In the former method, the density is visualized by adding phenol red, while in the CsCl purification, AAV particles can be seen as viral capsid band by naked eyes but require large-scale transfection (transfection of 60 × 15 cm dishes).

4.4.1 Iodixanol (OptiPrep) Purification

1. Combine cell-associated (Subheading 4.3.1) and media-secreted AAV (Subheading 4.3.2).
2. Prepare gradient solution (*see* Table 1 for gradients).
3. Deliver each gradient solution to the bottom of the centrifuge tubes starting from low density and finishing with high density (*see* Table 2 for volume). Start by adding the 15% gradient, and then carefully underlay the 25% gradient. Continue to underlay the higher-density gradient.
4. Load the combined AAV above the 15% gradient.
5. Centrifuge at 238,000 × *g* for 60 min at 20 °C for a NVT65 rotor (*see* **Note 10**).
6. Bring the tubes to the tissue culture hood, and place into the holder.
7. Wipe the side of the tube with 70% ethanol.
8. Remove the cap.
9. Insert a 23 g needle on a 3 mL syringe below the 40/60 interface.
10. Collect 1.0–1.5 mL from the 40–60% interface, not collecting any of the visible band at the 20/40% interface (*see* **Note 11**).
11. Transfer the virus to a 50 mL tube, and add DPBS to 50 mL.

Table 1
Preparation of iodixanol gradient

Percentage of iodixanol	OptiPrep (60%) (mL)	PBS-MK (mL)	Phenol red (μL)
15% iodixanol	4	12	
25% iodixanol	6.7	9.3	40
40% iodixanol	10	5	
60% iodixanol	10		25

Diluted iodixanol can be stored at 4 °C for at least 1 week.

Table 2
Set-up gradient

	Beckman #362181 (SW32 Ti) (mL)	OptiSeal #362181 (NVT65) (mL)
15% iodixanol	7	1.5
25% iodixanol	5	1.3
40% iodixanol	5	1.4
60% iodixanol	4	1.3
Lysate	~17	~5
Total volume	38.5	10.5
Speed	25k	50k
Centrifugation time	15 h	1 h

12. Equilibrate spin column (Amicon Ultra-15 Centrifugal Filter) with 15 mL of DPBS by centrifuging at 1430 × *g* for 5 min.
13. Load with diluted AAV in batches, and centrifuge at 1430 × *g* for 5 min several times (discard filtrate between centrifugation) until all virus solution has been applied.
14. Wash the filter by adding 15 mL of DPBS, and spin through twice.
15. Centrifuge until approximately 200 μL remains in the last wash.
16. Transfer to a new 0.5 mL tube, rinse membrane of the spin column with 50 μL of DPBS, and combine.
17. AAV is stable at least for 6 months at 4 °C. For longer storage, store at −80 °C.

4.4.2 CsCl Purification

1. Combine cell-associated (Subheading 4.3.1) and media-secreted AAV (Subheading 4.3.2).
2. Set up gradient.

3. For NVT65 OptiSeal tube, deliver 3.5 mL of the low-density CsCl solution (1.3 g/mL) to the bottom of the centrifuge tubes, and underlay 1.75 mL of the high-density CsCl solution (1.5 g/mL). Mark the density by Sharpie.
4. Overlay 5 mL of viral suspension to the top.
5. Centrifuge in NVT65 rotor at 238,000 × *g* for 2.5 h at 20 °C.
6. Clean the area on the tube where AAV presents and collects viral band using 3 mL syringe with 23 g needle (*see* **Note 12**).
7. Transfer AAV to a new centrifuge tube. Fill the top with low-density CsCl (1.3 g/mL), and centrifuge in NVT65 rotor at 238,000 × *g* for 16 h at 20 °C.
8. Collect the viral band using 3 mL syringe with 23 g needle.
9. Transfer to a dialysis cassette, and dialyze against PBS containing 5% sorbitol and 0.1% Tween 80 for 4 h.

4.5 Determination of AAV Titers by the Quantitative PCR

Quantitative PCR (QPCR) is the standard method to quantify vector genomic DNA which is proportional to number of AAV particles.

4.5.1 Physical Titer

1. Prepare serial dilution of standard stock solution (100 ng/mL) to cover 0.1–0.00001 ng/mL (*see* **Note 13**).
2. Prepare serial dilution of AAV starting at 1:100 and ending at 1:100,000 (*see* **Note 14**).
3. Prepare QPCR mix for projected numbers of wells (add 10 μL of 2× SYBR Green Master Mix, 1 μL of 10 pmol/mL solution of each forward and reverse primer, and 3 μL of H_2O/well) (*see* **Note 15**).
4. Pipette 15 μL of PCR/primer mix to each well of 96-well PCR plate in duplicate.
5. Pipette 5 μL of diluted standard and diluted samples in duplicate.
6. Place the lid over each well, and centrifuge at 740 × *g* for 2 min to remove any air bubbles.
7. Run QPCR.
8. Calculate titer using the following formula with 6614 bp AAV transfer vector as standard.

   Titer = vector genome copy/μL = 1×10^9 vector genome copy × (concentration in ng/μL from Q-PCR) × (dilution factor)/3.6245 ng (*see* **Note 16**).

4.5.2 Infectious Titer

Infectious titer is influenced by serotype and expression of its target cell attachment factors. However, the method described below determines the relative infectivity of AAV vectors:

1. Seed 1×10^5 HEK293T cells in 0.5 mL of DMEM supplemented with 5% FBS in a 24-well plate.
2. Add 2 μL of undiluted AAV in duplicate. Control wells have no infection.
3. Incubate at 37 °C for 24 h in humidified CO_2 incubator.
4. Trypsinize cells in control wells, and count number of cells/well (*see* **Note 17**).
5. Harvest AAV-infected cells, and transfer to a 1.5 mL microcentrifuge tube.
6. Extract DNA using a kit, and elute with 100 μL.
7. Measure DNA concentration by OD_{260nm}.
8. Quantify AAV vector genome using 5 μL of undiluted sample as described in Subheading 4.5.1.
9. Calculate the infectious titer using the following formula.

   Transducing unit (TU)/μL = (copy number/genome) × (number of cells 24 h after infection) × 1/2 μL (2 μL vector was used to infect) = [concentration of AAV vector in ng/μL (from Q-PCR)/(1.10233^{-6} ng × DNA concentration in ng/μL)] × 0.5 × (number of cells 24 h after infection).

5 Notes

1. The yield of AAV2 is low compared to other serotypes. Aggregation of AAV2 particles has been implicated. An isotonic formulation with elevated ionic strength has been reported to prevent AAV2 aggregation [17].
2. The solution is extremely viscous and will take several hours to filter. It takes time to dissolve PEG solution. Therefore, prepare ahead of time.
3. In our experience, Clontech In-Fusion Cloning Kit with CloneAmp HiFi PCR Premix or GenScript GenBuilder DNA Assembly with Clontech CloneAmp HiFi PCR Premix works well.
4. The doubling time of HEK293T cells differs from source to source and should be determined in each lab. A rule of thumb is that 1:5.5 to 1:6 split of 70–80% confluent cells (~4.4×10^6 cells/15 cm dish) reaches 70–80% confluency 24 h after split.
5. Transfection efficiency should be over 80%.
6. Cell suspension can be stored at −80 °C. For AAV2, use CSB2 instead of CSB.
7. AAV containing supernatant can be stored at −80 °C. This is a good break point.

8. AAV in PEG solution is stable for at least 1 week.
9. It takes time to dissolve the pellets. Vortex the tube, mix by pipetting up and down, and leave at room temperature. Make sure that there are no visible pellets.
10. For a SW32 Ti rotor, centrifuge at 59,400 $\times$ *g* for 16 h at 20 °C.
11. There should be no visible virus band corresponding to particles containing AAV genome.
12. There should be two bands. The top band is empty particles, and the bottom band is particles containing AAV genome. If a titer is low, it is difficult to see a band. Use bright light under the tube to visualize the band. To see a band, at least 1×10^{13} vector genome is necessary.
13. Dilutions can be made at copy number/mL; however, DNA concentrations are more convenient for calculation. Linearized plasmids are often used as standard in combination with the PCR primers targeting outside the ITR [18]. Circular plasmids are also good as standards as long as PCR primers do not target ITR.
14. Higher dilutions (1:1000 to 1:10,000) of most AAV preparations fall into the linear range of standard.
15. PCR primers should target expression cassettes with the size of PCR product between 150 and 250 bp. Primer can be designed using program such as Primer-BLAST (https://www.ncbi.nlm.nih.gov/tools/primer-blast/index.cgi?LINK_LOC=BlastHome).
16. AAV is a single-stranded DNA virus. If the standard AAV transfer vector genome size is 6614 bp, 1×10^9 vector genome = 6614 bp $\times$ 660 (Da/bp) $\times$ 1×10^9/Avogadro's number (6.022×10^{23}) $\times$ 1/2 = 3.6245 ng.
17. Most AAV does not integrate, and cell division dilutes vector genome/cell. Therefore, total number of cells in a well must be counted at the time of harvest. Transfer remaining cells to a 1.5 mL micro-centrifuge tube for extraction of cellular DNA.

Acknowledgment

We thank Dr. Sean Hartig for helpful discussion and critical reading of the manuscript. This work was supported by Gene Vector Core, Baylor College of Medicine Advanced Technology Cores.

References

1. Atchison RW, Casto BC, Hammon WM (1965) Adenovirus-associated defective virus particles. Science 149(3685):754–756
2. Pillay S, Carette JE (2017) Host determinants of adeno-associated viral vector entry. Curr Opin Virol 24:124–131. https://doi.org/10.1016/j.coviro.2017.06.003
3. Dalkara D, Byrne LC, Klimczak RR, Visel M, Yin L, Merigan WH, Flannery JG, Schaffer DV (2013) In vivo-directed evolution of a new adeno-associated virus for therapeutic outer retinal gene delivery from the vitreous. Sci Transl Med 5(189):189ra176. https://doi.org/10.1126/scitranslmed.3005708
4. Ramachandran PS, Lee V, Wei Z, Song JY, Casal G, Cronin T, Willett K, Huckfeldt R, Morgan JI, Aleman TS, Maguire AM, Bennett J (2017) Evaluation of dose and safety of AAV7m8 and AAV8BP2 in the non-human primate retina. Hum Gene Ther 28 (2):154–167. https://doi.org/10.1089/hum.2016.111
5. Grimm D, Zolotukhin S (2015) E pluribus Unum: 50 years of research, millions of viruses, and one goal—tailored acceleration of AAV evolution. Mol Ther 23(12):1819–1831. https://doi.org/10.1038/mt.2015.173
6. Weinmann J, Grimm D (2017) Next-generation AAV vectors for clinical use: an ever-accelerating race. Virus Genes 53 (5):707–713. https://doi.org/10.1007/s11262-017-1502-7
7. Colella P, Ronzitti G, Mingozzi F (2018) Emerging issues in AAV-mediated in vivo gene therapy. Mol Ther Methods Clin Dev 8:87–104. https://doi.org/10.1016/j.omtm.2017.11.007
8. Nakai H, Yant SR, Storm TA, Fuess S, Meuse L, Kay MA (2001) Extrachromosomal recombinant adeno-associated virus vector genomes are primarily responsible for stable liver transduction in vivo. J Virol 75 (15):6969–6976. https://doi.org/10.1128/JVI.75.15.6969-6976.2001
9. Bell P, Moscioni AD, McCarter RJ, Wu D, Gao G, Hoang A, Sanmiguel JC, Sun X, Wivel NA, Raper SE, Furth EE, Batshaw ML, Wilson JM (2006) Analysis of tumors arising in male B6C3F1 mice with and without AAV vector delivery to liver. Mol Ther 14(1):34–44. https://doi.org/10.1016/j.ymthe.2006.03.008
10. Zhong L, Malani N, Li M, Brady T, Xie J, Bell P, Li S, Jones H, Wilson JM, Flotte TR, Bushman FD, Gao G (2013) Recombinant adeno-associated virus integration sites in murine liver after ornithine transcarbamylase gene correction. Hum Gene Ther 24 (5):520–525. https://doi.org/10.1089/hum.2012.112
11. Nault JC, Datta S, Imbeaud S, Franconi A, Mallet M, Couchy G, Letouze E, Pilati C, Verret B, Blanc JF, Balabaud C, Calderaro J, Laurent A, Letexier M, Bioulac-Sage P, Calvo F, Zucman-Rossi J (2015) Recurrent AAV2-related insertional mutagenesis in human hepatocellular carcinomas. Nat Genet 47(10):1187–1193. https://doi.org/10.1038/ng.3389
12. Dong B, Nakai H, Xiao W (2010) Characterization of genome integrity for oversized recombinant AAV vector. Mol Ther 18 (1):87–92. https://doi.org/10.1038/mt.2009.258
13. Ghosh A, Duan D (2007) Expanding adeno-associated viral vector capacity: a tale of two vectors. Biotechnol Genet Eng Rev 24:165–177
14. Maddalena A, Tornabene P, Tiberi P, Minopoli R, Manfredi A, Mutarelli M, Rossi S, Simonelli F, Naggert JK, Cacchiarelli D, Auricchio A (2018) Triple vectors expand AAV transfer capacity in the retina. Mol Ther 26(2):524–541. https://doi.org/10.1016/j.ymthe.2017.11.019
15. Ayuso E, Mingozzi F, Montane J, Leon X, Anguela XM, Haurigot V, Edmonson SA, Africa L, Zhou S, High KA, Bosch F, Wright JF (2010) High AAV vector purity results in serotype- and tissue-independent enhancement of transduction efficiency. Gene Ther 17 (4):503–510. https://doi.org/10.1038/gt.2009.157
16. Lock M, Alvira M, Vandenberghe LH, Samanta A, Toelen J, Debyser Z, Wilson JM (2010) Rapid, simple, and versatile manufacturing of recombinant adeno-associated viral vectors at scale. Hum Gene Ther 21(10):1259–1271. https://doi.org/10.1089/hum.2010.055
17. Wright JF, Le T, Prado J, Bahr-Davidson J, Smith PH, Zhen Z, Sommer JM, Pierce GF, Qu G (2005) Identification of factors that contribute to recombinant AAV2 particle aggregation and methods to prevent its occurrence during vector purification and formulation. Mol Ther 12(1):171–178. https://doi.org/10.1016/j.ymthe.2005.02.021
18. D'Costa S, Blouin V, Broucque F, Penaud-Budloo M, Francois A, Perez IC, Le Bec C, Moullier P, Snyder RO, Ayuso E (2016) Practical utilization of recombinant AAV vector reference standards: focus on vector genomes titration by free ITR qPCR. Mol Ther Methods Clin Dev 5:16019. https://doi.org/10.1038/mtm.2016.19

Chapter 8

Transposon-Mediated Stable Suppression of Gene Expression in the Developing Chick Retina

Masaru Nakamoto and Chizu Nakamoto

Abstract

The embryonic chick has long been a favorite model system for in vivo studies of vertebrate development. However, a major technical limitation of the chick embryo has been the lack of efficient loss-of-function approaches for analyses of gene functions. Here, we describe a methodology in which a transgene encoding artificial microRNA sequences is introduced into embryonic chick retinal cells by in ovo electroporation and integrated into the genome using the Tol2 transposon system. We show that this methodology can induce potent and stable suppression of gene expression. This technique therefore provides a rapid and robust loss-of-function approach for studies of gene function in the developing retina.

Key words In ovo electroporation, Chick embryo, Retina, Tol2 transposon, miRNA, Gene targeting

1 Introduction

The chick embryo has long been a remarkably important model system in developmental biology. In the field of neural development, the chick retina has contributed to considerable progress in understanding the molecular mechanisms by which cell proliferation and differentiation, morphogenesis, and axon guidance are regulated. The chick retina has significant similarity to the human retina at the anatomical, cellular, and molecular levels. In addition, the practical advantages of the chick retina include its large size, rapid development, and accessibility for visualization and experimental manipulation. The usefulness of the chick embryo as an experimental tool has recently been further expanded by invention and development of the in ovo electroporation technique by Nakamura's group [1, 2].

Despite these advantages, a major technical limitation of the chick embryo has been the lack of a technology that allows for robust and efficient loss-of-function analyses, which have been dependent on the availability of a dominant-negative form of the gene product. Nonetheless, important advances have been made in

Chai-An Mao (ed.), *Retinal Development: Methods and Protocols*, Methods in Molecular Biology, vol. 2092,
https://doi.org/10.1007/978-1-0716-0175-4_8,

this regard in the past decades. For example, RNAi has been successfully used in chick embryos by electroporating small interfering RNAs (siRNAs) [3] or expression vectors for short hairpin RNA [4] in ovo. In addition, an RCAS retrovirus-based siRNA delivery system was shown to efficiently knock down gene expression in chick embryos [5].

Transposons are genetic elements that move from one locus in the genome to another and can be efficient tools for permanently delivering foreign DNA into vertebrate genomes. The Tol2 transposable element, originally identified in the genome of the medaka fish (*Oryzias latipes*), is an autonomous transposon that encodes a functional transposase belonging to the hAT family [6]. The Tol2 transposase can catalyze transposition of a nonautonomous transposon construct, which lacks the transposase coding region but retains the sequences of the left and right ends of the Tol2 element (200 bp and 150 bp, respectively). When a plasmid vector for a nonautonomous transposon construct carrying a gene expression cassette with the Tol2 end sequences is introduced into vertebrate cells with the Tol2 transposase activity, the transposon construct is excised from the plasmid, and the expression cassette is integrated into the host genome (Fig. 1). It has been reported that the Tol2 transposable element can undergo efficient transposition in a wide variety of vertebrate species, including zebrafish, frogs, chick, and mice [7–11].

In this chapter, we describe a loss-of-function approach in which artificial microRNAs (miRNAs) are introduced into the developing chick retina by in ovo electroporation and subsequently integrated into the genome using a Tol2 transposon system [12]. The outline of the procedures is shown in Fig. 2. In this approach, first, DNA oligonucleotides encoding the target pre-miRNA and its complement are designed and generated (Subheading 3.1). The complementary oligonucleotides are then annealed to generate double-stranded oligonucleotides (Subheading 3.2) and cloned into an expression vector for miRNA and the EmGFP marker (Subheading 3.3). Knockdown effects of individual miRNAs can be tested in vitro (Subheading 3.4), and miRNA sequences that show strong activities can be chained to further enhance the knockdown effects (Subheading 3.5). The expression cassette carrying EmGFP cDNA and pre-miRNA sequences is then transferred to the Tol2 transposon vector (Subheading 3.6). Finally, the Tol2 transposon construct is introduced with the transposase expression vector into the embryonic chick retina by in ovo electroporation (Subheadings 3.7–3.9).

This method offers several distinct advantages. First, it supports stable expression of the transgene and thus persistent production of miRNAs. Electroporation of conventional expression vectors permits only transient expression of the transgene, usually lasting only for several days in embryonic cells, because of the inability of the

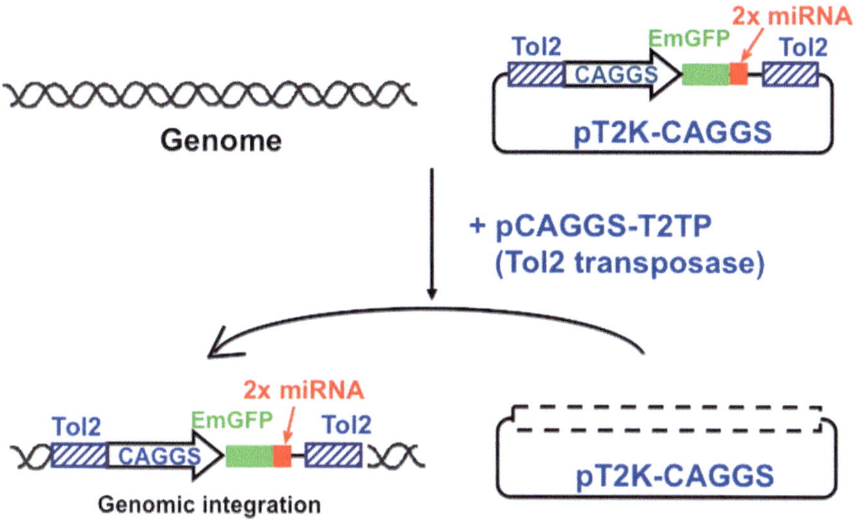

Fig. 1 Transposition of Tol2-flanked DNA by transposase. A scheme showing transposition of a Tol2-flanked expression cassette for EmGFP and artificial miRNAs (two miRNA sequences are chained) by transposase. When a Tol2 transposon construct containing the expression cassette (pT2K-CAGGS-EmGFP/miRNA) is introduced into cells with a Tol2 transposase expression construct (pCAGGS-T2TP), the Tol2-flanked cassette is excised from the vector and integrated into the host genome by the transposase activity. Expression of EmGFP and miRNAs is driven by the ubiquitous promoter CAGGS (modified from [11, 12])

electroporated plasmid to be integrated into chromosomes. In contrast, the present method involves chromosomal integration of the expression cassette by activity of co-electroporated transposase and therefore can achieve stable expression of the transgenes. Second, the cells that express the transgene can be easily traced because each of the transfected cells is clearly marked by the EmGFP marker. Third, because expression of miRNA and EmGFP is driven by the same CAGGS promoter, the risk of promoter interference is avoided, which is one of the common problems with retroviral vectors containing an additional promoter to attain dual gene expression (e.g., [13]). Finally, the areas of transfection can be controlled by properly choosing the type and positions of electrodes. Because this method does not use replication-competent retroviral vectors, there is no transmission of the transgene to neighboring cells after the electroporation. Within the transfected retinae, areas that do not express the transgene serve as a good internal control. Therefore, this system provides a stable and efficient loss-of-function option and can contribute to the analysis of gene functions during chick development.

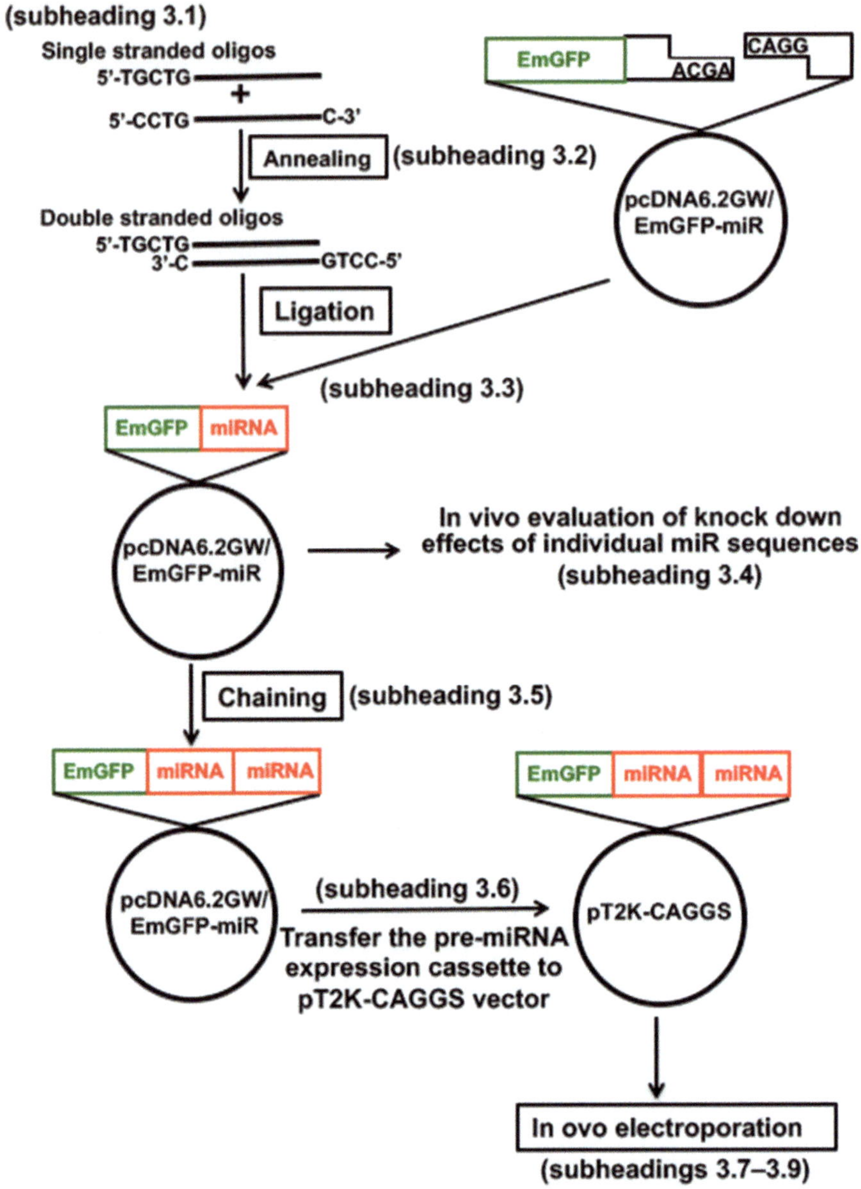

Fig. 2 Outline of the procedures. This diagram shows key steps of the procedures. DNA oligonucleotides encoding the target pre-miRNA and its complement are designed and generated (Subheading 3.1). The complementary oligonucleotides are annealed to generate double-stranded oligonucleotides (Subheading 3.2). Individual double-stranded oligos are cloned into an expression vector for miRNA and the EmGFP marker (Subheading 3.3). Knockdown effects of individual miRNA can be tested in vitro (Subheading 3.4), and miRNA sequences that show strong activities can be chained to further enhance the knockdown effects (Subheading 3.5). The expression cassette carrying EmGFP cDNA and pre-miRNA sequences is then transferred to the Tol2 transposon vector (Subheading 3.6). Finally, the Tol2 transposon construct is introduced with the transposase expression vector into the embryonic chick retina by in ovo electroporation (Subheadings 3.7–3.9)

2 Materials

2.1 Construction of miRNA Expression Vectors

1. *RNAi Designer* (https://rnaidesigner.thermofisher.com/rnaiexpress/). An online tool to choose target sequences and design pre-miRNA sequences
2. The BLOCK-iT Pol II miR RNA expression kit with EmGFP (Invitrogen) https://www.thermofisher.com/order/catalog/product/K493600?SID=srch-hj-K4936-00

   The kit contains the following reagents to be used in the procedures:

   (a) pcDNA6.2-GW/EmGFP-miR vector, linearized

   (b) 10× Oligo annealing buffer: 100 mM Tris–HCl (pH 8.0), 10 mM EDTA (pH 8.0), and 1 M NaCl

   (c) DNase/RNase-free water

   (d) 5× Ligation buffer: 250 mM Tris–HCl (pH 7.6), 50 mM $MgCl_2$, 5 mM ATP, 5 mM DTT, and 25% (w/v) polyethylene glycol-8000

   (e) T4 DNA ligase

   (f) EmGFP forward sequencing primer: 5′-GGCATGGACGAGCTGTACAA-3′

   (g) miRNA reverse sequencing primer: 5′-CTCTAGATCAACCACTTTGT-3′

   (h) pcDNA6.2-GW/miR-negative control plasmid

   (i) S.O.C medium: 2% tryptone, 0.5% yeast extract, 10 mM NaCl, 2.5 mM KCl, 10 mM $MgCl_2$, 10 mM $MgSO_4$, and 20 mM glucose

   (j) One Shot TOP10 chemically competent *E. coli*
3. Construction of Tol2 transposon vectors

   Tol2 transposon vectors:

   (a) pT2K-CAGGS vector (Fig. 4): kindly provided by Yoshiko Takahashi (Kyoto University, Japan)

   (b) pCAGGS-T2TP vector: T2 transposase expression plasmid containing the transposase cDNA under the control of the CAGGS promoter. A generous kind gift of Koichi Kawakami (National Institute of Genetics, Japan). Also available from Addgene.

   Primers:

   (a) Forward primer (the 3′ end region of the CMV promoter of the pcDNA6.2-GW/EmGFP-miR vector with an artificial EcoRI site): 5′-GGGAATTCTCTGGCTAACTAGAGAAC-3’

   (b) Reverse primer (miRNA reverse sequencing primer with an artificial EcoRI site): 5′-CCGAATTCCCTCTAGATCAACCACT-3’

4. Spectinomycin
5. Thermal cycler

2.2 In Vitro Evaluation of Knockdown Effects Using Alkaline Phosphatase Fusion Proteins

1. APtag vectors:
   Different types of vectors to make alkaline phosphatase (AP) fusion proteins (APtag-1–APtag-5) are commercially available from GenHunter Corp (TN, USA) https://www.genhunter.com/products/ap-tag-kit-a.html. We generally use the APtag-2 vector, which is designed for transient transfection and for fusions to the N-terminus of AP.
2. 2× AP substrate buffer:

10 M Diethanolamine (pH 9.8)	4 mL
1 M $MgCl_2$	10 μL
L-Homoarginine	45 mg
BSA	10 mg
p-Nitrophenylphosphate	63 mg
Add dH_2O to total volume of 20 mL.	
The solution can be stored in aliquots at −20 °C. All reagents can be obtained from Sigma-Aldrich (St. Louis, MO)	

3. Microplate reader or spectrophotometer

2.3 Preparation of Chick Embryos

1. Fertilized chicken eggs
   Fertilized chicken eggs (*Gallus gallus*) can be obtained from local farms. Alternatively, pathogen-free chicken eggs can be purchased from commercial vendors (e.g., Charles River). The eggs can be stored at 12–16 °C for up to 1 week before use (*see* **Note 1**).
2. Incubation of chick embryos
   Chicken eggs are incubated at 38 °C with sufficient humidity. A variety of incubators are available, such as Kuhl—600 Egg Research and Laboratory Incubator (B-Lab-600-110) (https://www.flemingoutdoors.com/kuhl%2D%2D-600-egg-laboratory-incubator%2D%2D-b-lab-600-110.html).

2.4 In Ovo Microinjection and Electroporation

Equipment for in ovo microinjection and electroporation is presented in Fig. 5:

1. Dissecting microscope
2. Gooseneck fiber light source
3. Microinjection apparatus:

Micromanipulator (e.g., Narishige (Japan), MM3), http://products.narishige-group.com/group1/MM-3/electro/english.html

18 G Needles

PVC tubing (VWR, Cat. No. 228-3830)

Hamilton syringe (50 μL, Sigma Cat. No. 20715, Hamilton Cat. No. 80901).

Glass micropipette needles:

Micropipette needles used for microinjection are made by pulling FHC borosilicate capillary tube with omega dot fiber (1 mm O.D. × 0.75 mm I.D., FHC Inc. ME, USA, Cat. No. 30-30-1) with a micropipette puller (e.g., Sutter Instrument P97, https://www.sutter.com/MICROPIPETTE/p-97.html). Pinch off the tip of pulled capillary tubes with fine forceps to generate an opening.

4. Electroporation apparatus:
   Electroporator—CUY21 Square Wave Electroporator (Nepa Gene)

   Electrodes (CUY611P3-1, Sonidel)

   https://www.sonidel.com/product_info.php?products_id=94

   Holder (CUY580, Sonidel)

   https://www.sonidel.com/product_info.php?products_id=85

   Cables:

   C117 cables, https://www.sonidel.com/product_info.php?products_id=252

   C115CB cables, https://www.sonidel.com/product_info.php?products_id=254
5. Forceps: fine, curved, and straight shape
   Scissors: medium and small size
6. Heavy mineral oil (Sigma-Aldrich).
7. 1% Fast Green Solution:
   100 mg of Fast Green FCF (Sigma-Aldrich) is dissolved in 10 mL of PBS (with Ca^{2+} and Mg^{2+}; pH 7.4). Filter the solution through 0.2 μm syringe filter into 15 mL conical tube. Store at room temperature.
8. Hank's saline solution

3 Methods

3.1 Designing Single-Stranded DNA Oligos for Targeting the Gene of Interest [14]

The design of pre-miRNA oligonucleotides is critical for successful gene silencing. We use an online tool provided by Invitrogen, *RNAi Designer*, which helps choose target sequences and design pre-miRNA oligos. For a given target gene, it may be required to generate multiple pre-miRNA sequences and screen them for

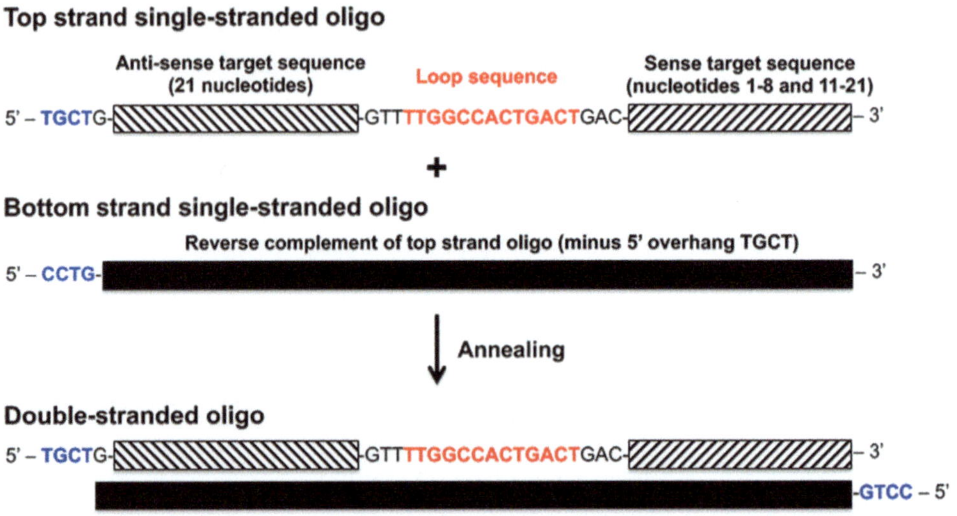

Fig. 3 Design of single-stranded oligonucleotides. The required features of the top- and bottom-strand single-stranded oligonucleotides are shown. The backbone of the miRNA sequence is derived from murine miR-155. The top-strand oligo contains (from 5′ to 3′) a sequence for 5′ overhang, antisense target sequence (mature miRNA sequence), terminal loop sequence (red), and sense target sequence. The bottom-strand oligo represents the reverse complement of the top-strand oligo with additional four nucleotide (CCTG) at the 5′ end and without the four nucleotides at the 3′ end (AGCA: reverse complement of 5′-TGCT of the top-strand oligo). 5′ overhangs are indicated in blue (modified from [14])

knockdown activities. We generally test 5–10 pre-miRNA sequences for knockdown effects in vitro (Subheading 3.4). We then select and chain two most efficient sequences (Subheading 3.5) for using in ovo electroporation.

For each pre-miRNA sequence, two single-stranded DNA oligos are required: one encoding the target pre-miRNA ("top-strand" oligo) and the other its complement ("bottom-strand" oligo) (Fig. 3).

3.1.1 Top Strand Oligo

The top-strand oligo contains the following features (from 5′ to 3′):

1. 5′ TGCTG: Derived from the endogenous murine miR-155, which is the basis of the pcDNA 6.2-GW/EmGFP-miR vector system [15]. This also provides a four-nucleotide 5′ overhang, compatible with a four-nucleotide overhang in the linearized pcDNA 6.2-GW/EmGFP-miR, to clone the double-stranded oligo.
2. Reverse complement of the 21-nucleotide sense target sequence.
3. GTTTTGGCCACTGACTGAC (terminal loop): Derived from miR-155 to form a terminal loop.

4. Nucleotides 1–8 and 11–21 (5′–3′) of sense target sequence. (The nucleotides 9 and 10 are removed to form a short internal loop in the mature miRNA, which results in more efficient knockdown [14].)

3.1.2 Bottom-Strand Oligo

The bottom-strand oligo can be designed as follows:

1. Remove 5′ TGCT from top oligo sequence (new sequence starts with G).
2. Take the reverse complement of the sequence from **step 1**.
3. Add CCTG to the 5′ end of the sequence from **step 2**. The four nucleotides are derived from endogenous miR-155. This also constitutes the four-nucleotide 5′ overhang, compatible with a four-nucleotide overhang in the provided linearized pcDNA6.2-GW/EmGFP-miR, to clone the double stranded oligo.

3.2 Annealing of the Top- and Bottom-Strand Oligos [14]

Anneal the top- and bottom-strand oligos to generate a double-stranded oligonucleotide suitable for cloning into the pcDNA6.2-GW/EmGFP-miR vector:

1. In a 0.5 mL sterile microcentrifuge tube, set up the following annealing reaction at room temperature:

Top-strand oligo (200 μM in TE buffer)	5 μL	(final concentration 50 μM)
Bottom-strand oligo (200 μM in TE buffer)	5 μL	(final concentration 50 μM)
10× Oligo annealing buffer	2 μL	
DNase/RNase-free water	8 μL	
Total volume	20 μL	

2. Incubate at 95 °C for 4 min.
3. Allow the reaction mixture to cool to room temperature for 5–10 min. The single-stranded oligos will anneal during this time.
4. Place the sample in a microcentrifuge, and centrifuge briefly (~5 s). Mix gently. The double-stranded oligo in the annealing mixture can be stored at −20 °C stable for at least a year.

3.3 Cloning the Double-Stranded Oligonucleotides into the pcDNA6.2-GW/EmGFP-miRNA Vector [14]

1. Dilute the double-stranded oligos (50 μM) to a final concentration of 10 nM (i.e., 5000-fold dilution) in two-step serial dilutions:

   (Step 1) Using 1 μL of the annealing mixture (from the previous section), dilute the 50 μM mixture 100-fold into DNase/RNase-free water to a final concentration of 500 nM. Vortex to mix thoroughly.

50 μM double-stranded oligo	1 μL
DNase/RNase-free water	99 μL
Total volume	100 μ L

(Step 2) Using 1 μL of the 500 nM double-stranded mixture (from Step 1), dilute 50-fold into the oligo annealing buffer to a final concentration of 10 nM. Vortex to mix thoroughly. Store the remaining 500 nM double-stranded oligo stock at −20 ° C.

500 nM double-stranded oligo	1 μL
10× oligo annealing buffer	5 μL
DNase/RNase-free water	44 μL
Total volume	50 μL

Aliquot the 10 nM double-stranded oligo stock, and store at −20 ° C.

2. Set up a 20 μL ligation reaction at room temperature using the following reagents *in the order* below:

5× Ligation buffer	4 μL
Linearized pcDNA6.2-GW/EmGFP-miR (5 ng/μL)	2 μL
Double-stranded oligo (10 nM in 1× oligo annealing buffer) (from the previous step)	4 μL
DNase/RNase-free water	9 μL
T4 DNA ligase (1 U/μL)	1 μL
Total	20 μL

3. Mix reaction thoroughly by pipetting up and down (do not vortex).
4. Incubate for 5 min at room temperature.
5. Place the reaction on ice. The ligation reaction can be stored at −20 °C overnight.
6. Add 2 μL of the ligation reaction into a vial of One Shot TOP10 chemically competent *E. coli* on ice, and mix gently by tapping. Do not mix by pipetting up and down.
7. Incubate on ice for 10 min (5–30 min).
8. Heat-shock the cells for 30 s at 42 °C, and immediately transfer the tubes on ice.

9. Add 250 μL of room-temperature S.O.C. medium, and incubate with shaking (200 rpm) at 37 °C for 1 h.
10. Spread 50–200 μL from each transformation on a pre-warmed LB agar plate containing 50 μg/mL spectinomycin, and incubate overnight at 37 °C.
11. Pick up 5–10 spectinomycin-resistant colonies, isolate plasmid DNA, and confirm the sequence and orientation of the double-stranded oligo insert using the EmGFP forward sequencing primer and miRNA reverse sequencing primer.

3.4 Evaluation of the Knockdown Effects of pcDNATM6.2-GW/EmGFP-miR Expression Clones

We generally suggest that gene suppression efficiency of individual miRNA sequences is evaluated in transient RNAi analysis in vitro, before applying the miRNA sequences in ovo. Knockdown efficiency can be evaluated by, for example, introducing the pcDNA6.2-GW/-EmGFP-miR plasmids into cell lines that are known to express endogenously the target gene.

Alternatively, individual pcDNA6.2-GW/-EmGFP-miR plasmids can be co-transfected into culture cells with an expression construct for the target gene. If the target gene encodes an extracellular protein (protein “X”), it can be fused with an alkaline phosphatase (AP) tag in the expression construct (“X”-AP fusion protein) by using an APtag vector [16], as used in our previous study [12]. After transfection, the “X”-AP fusion protein will be secreted to the culture media, and its expression levels and knockdown effects of miRNAs can be monitored by measuring the AP activities in the culture media:

1. Plate HEK293T cells in a 24-well plate (8×10^4 cells/well), and culture overnight at 37 °C in a 5% CO_2 incubator.
2. Next day, transfect individual pcDNA6.2-GW/EmFFP-miR constructs with an expression plasmid of an AP-tagged protein, by using a conventional transfection method.
3. 48–72 h After the transfection, collect the culture medium.
4. Put 1 mL of culture medium collected from each well in a 1.5 mL Eppendorf tube, and incubate in a 65 °C water bath for 5 min to heat inactivate endogenous APs.
5. Spin the tubes in a microcentrifuge at maximum speed for 5 min. Collect the supernatant.
6. Take 100 μL of the supernatant, and add an equal amount of 2× AP buffer to check the AP activity by measuring OD405 using a plate reader (*see* **Note 2**). If the activity is high, it may be necessary to dilute the supernatant first, using HBAH or another buffer containing a carrier protein (*see* **Note 3**).

3.5 Chaining of miRNA Sequence [14]

miRNAs are sometimes expressed in clusters in long primary transcripts driven by RNA Pol II [17]. The pcDNA6.2-GW/EmGFP-miR vector supports chaining of miRNAs to ensure co-cistronic expression of multiple miRNAs. Chaining different miRNAs targeting the same gene or repeating one miRNA can enhance knockdown effects. We usually chain two miRNAs targeting our gene of interest that show highest knockdown activities in vitro (Subheading 3.4) as follows (in this example, miR-1 sequence is cloned into the vector expressing miR-2):

1. Digest pcDNA6.2-GW/EmGFP-miR-1 (insert) with Bam H I and XhoI and pcDNA 6.2-GW/EmGFP-miR-2 (backbone) with Bgl II and XhoI.
2. Run the enzyme-digested insert and backbone on 2% agarose gel, excise the fragments, and purify them using a conventional gel extraction method (e.g., by using QIAEXII gene extraction kit).
3. Ligate the purified backbone and insert fragment at a 1:4 molar ratio, using T4 DNA ligase.

   Transform competent *E. coli* (e.g., One Shot TOP10), and analyze resulting clones as described in Subheading 3.3.
4. Test the chained construct for knockdown effects using in vitro assays as described in Subheading 3.4.

3.6 Transferring the EmGFP/Pre-miRNA Expression Cassette to the pTol2 Vector

Finally, the expression cassette containing an EmGFP cDNA and 2× pre-miR sequences is transferred to the pT2K-CAGGS vector (Fig. 4). The expression cassette can be amplified by PCR using primers with an artificial restriction enzyme site (e.g., EcoRI) and cloned into the pT2K-CAGGS vector. We generally PCR-amplify

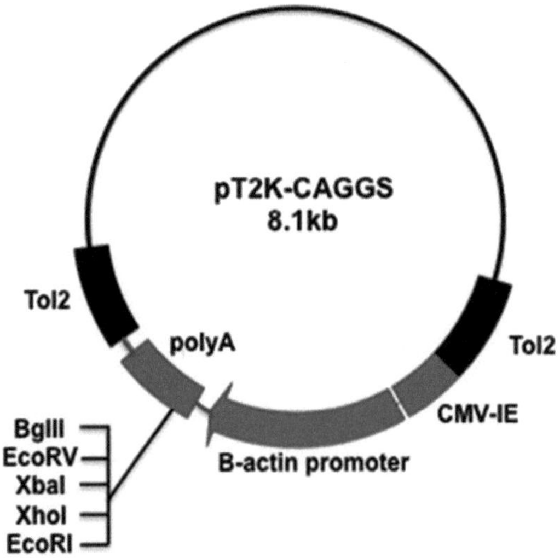

Fig. 4 The pT2K-CAGGS vector. Expression of ectopic genes inserted into the multicloning site is driven by the ubiquitous promoter CAGGS (including sequences from the cytomegalovirus intermediate early enhancer (CMV-IE) and chicken β-actin promoter). The expression cassette is flanked by the Tol2 sequences

and transfer the expression cassette encompassing from the 3′ end of the CMV promoter to the miRNA reverse sequencing primer site:

1. Conduct PCR as follows:

(Reaction setup)	
pcDNA6.2-GW/EmGFP/2x pre-miRNA plasmid or pcDNA6.2-GW/EmGFP-miRNA-negative control plasmid (0.1 μg/μL)	1 μL
10× PFU buffer	5 μL
dNTP mix (10 mM each of dATP, dCTP, dGTP, and dTTP)	1 μL
Forward primer (10 μM)	1 μL
Reverse primer (10 μM)	1 μL
Pfu DNA polymerase (five units/μL)	0.5 μL
dH_2O	40.5 μL
Total volume	50 μL

(Themocycling conditions)	
94 °C	5 mi
30 cycles of the following:	
94 °C	45 s
58 °C	1 min
72 °C	2 min
72 °C	10 min

2. Gel-purify the PCR product (c.1.3 kb), digest with the restriction enzyme (EcoRI), and ligate to the enzyme-digested pT2K-CAGGS vector (pT2K-CAGGS-EmGFP/miRNA). Transform *E. coli* (e.g., ElectroMAX DH10B) cells with the ligated plasmid, culture them overnight, and purify the plasmid by using a conventional Maxiprep method. Confirm the sequence of the expression cassette using the primers used for the PCR.

3.7 Preparation for In Ovo Microinjection and Electroporation

3.7.1 Chick Embryos

For in ovo microinjection and electroporation into the retina, we generally use chick embryos at HH stages 10–11 (HH10: 33–38 h, HH11: 40–45 h). The chicken eggs are incubated at 38 °C, with placing the eggs horizontally (the embryo will be positioned at the top of the egg). Label the eggs with the start date of incubation and their orientation in the incubator (e.g. mark the top side of the egg).

3.7.2 Setting Up of the Microinjection and Electroporation Apparatus

1. DNA cocktail
   DNA cocktails are made by mixing the individual pT2K-CAGGS-EmGFP/miRNA plasmids with the pCAGGS-T2TP plasmid (5 μg/μL each) at the ratio of 2:1.
2. Microinjection apparatus (Fig. 5a, b)
   The microinjection apparatus consists of (in this order) Hamilton syringe, 18 G needle, PCV tubing (c. 2 cm length), and microglass needle. To set up the apparatus, first, fill the inner space of the Hamilton syringe with heavy mineral oil. Then, attach an 18 G needle, and push the plunger of the Hamilton syringe to fill the inner space of the needle with heavy mineral oil. Repeat the same procedure for the PCV tubing and microglass needle. The tip of the microglass needle must be pinched off by fine forceps to make an opening. Avoid that air bubbles are trapped in the system. Finally, load DNA cocktail into the microglass needle.

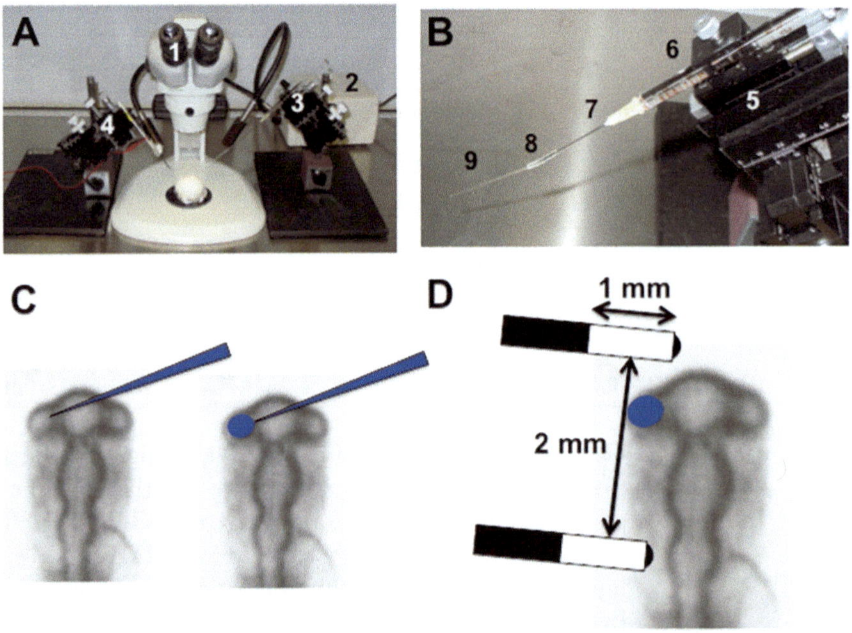

Fig. 5 In ovo microinjection and electroporation. (**a**) Typical work station for in ovo microinjection and electroporation. (*1*) Dissecting microscope. (*2*) Light source. (*3*) Injection apparatus set on a micromanipulator. (*4*) Electroporation apparatus set on a micromanipulator (electrodes are connected by cables to an electroporator (not shown)). (**b**) Injection apparatus set on a micromanipulator (*5*). A pulled glass needle (*9*) is connected to a Hamilton syringe (*6*) via PVC tubing (*8*) and an 18 G needle (*7*). The inner space of the apparatus is filled with heavy mineral oil. (**c**) Microinjection into the optic vesicle. DNA cocktail solution with fast green (blue) is injected into the optic vesicle with a micropipette needle. We usually insert the micropipette needle from the medial side of the optic vesicle. DNA solution is injected until the dye is visible throughout the optic vesicle. (**d**) Electroporation into the retina. Parallel platinum-coated electrodes are placed perpendicular to the long axis of the embryo: One electrode is placed near the anterior (nasal) optic vesicle, and the other is lateral to the hindbrain (posterior (temporal) side of the optic vesicle). The distance between the electrodes is 2 mm

3. Electroporation apparatus

   Set a pair of electrodes with an electrode holder on a micromanipulator (the distance between the electrodes is 2 mm), and connect the electrodes to an electroporator with cables (Fig. 5a).

3.8 Microinjection of Plasmid DNA Solution

1. Take out a chicken egg from the incubator, and place it on an egg holder horizontally (in the same orientation as during incubation).
2. Attach a sterile 18 G needle to a 10 mL syringe. Carefully insert the needle through the shell at the blunt end of the egg at a sharp downward angle (be careful not to damage the yolk).
3. Withdraw 3–5 mL of albumin (egg white) from the egg. This will make the embryo and vitelline membrane separate from the shell and thus avoid injury of the embryo or blood vessels when opening a window.
4. Cover the hole generated by the needle with Scotch tape.
5. Confirm that the embryo is detached from the shell, by "candling" the egg with light.
6. Open a window by removing an approximate 2–3 cm-diameter circle of shell from the top of the egg using forceps and scissors (*see* **Note 4**). Window only up to five eggs at any one time to prevent drying out of embryos during electroporation. A drop of Hank's solution on the embryo avoids drying of the embryo.
7. Insert the needle into the lumen of the optic vesicle, and inject DNA solution (with fast green) by slowly pushing (or tapping on) the plunger, until the dye is visible throughout the optic vesicle (Fig. 5c) (*see* **Note 5**). Remove the needle.

3.9 Electroporation

1. Place the electrodes on the vitelline membrane, perpendicular to the long axis of the embryo: one electrode being on the anterior (nasal) side and the other electrode on the posterior (temporal) side of the optic vesicle (lateral to the hindbrain). Position electrodes c. 2 mm apart (Fig. 5d). A drop of Hanks' solution could prevent damages to the embryo caused by the electroporation.
2. Apply five square pulses (voltage: 15 V, pulse length: 50 ms, pulse interval: 950 ms, pulse number: 5) by the CUY21 electroporator.
3. Remove the electrodes. Seal the window with Scotch tape. Re-incubate the embryos at 38 °C until analyses (Fig. 6) (*see* **Notes 6** and 7).

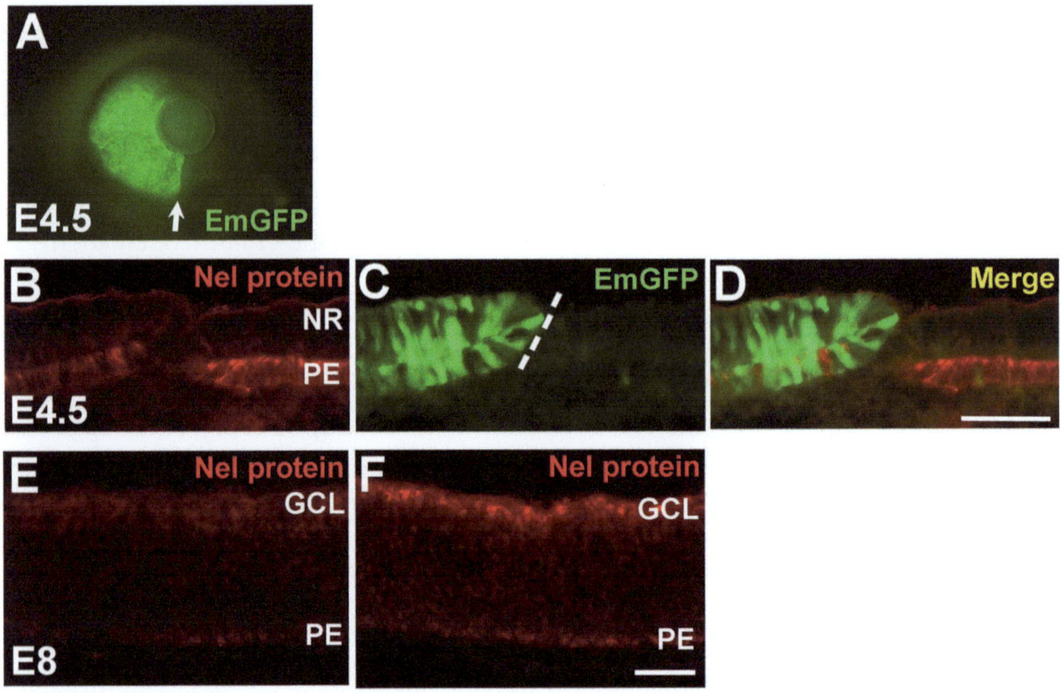

Fig. 6 Stable suppression of gene expression by Tol2 transposon-mediated integration of miRNA transgene. Nel (also known as Nell2) is an extracellular glycoprotein that regulates differentiation, survival, and axon guidance of retinal ganglion cells. During retinal development, Nel is expressed in the pigment epithelium and retinal ganglion cells. A transposon construct containing an expression cassette for EmGFP and miRNA sequences against the Nel gene was co-transfected with a transposase expression vector into the chick retina by in ovo electroporation at HH9-11. (**a**) Half of the retina (nasal or temporal side) was transfected with the transposon constructs and marked by EmGFP (green) (an example of transfection into the temporal retina is presented). The optic fissure (white arrow) represents the boundary between transfected and untransfected (control) sides. (**b**–**f**) Retinal sections were prepared at E4.5 (**b**–**d**) or E8 (**e**, **f**), and expression of Nel was examined by immunohistochemistry using anti-Nel antibody (red). Nel expression is significantly reduced on the transfection side (in the pigment epithelium (PE) at E4.5 and in the PE and ganglion cell layer (GCL) at E8). Note the complementary pattern of Nel immunostaining and EmGFP expression (**b**–**d**). Dotted line indicates the position of the optic fissure (modified from [12])

4 Notes

1. Alternatively, the eggs can be stored at 4 °C for up to 1 week. Extended storage would result in decreased rates of embryonic survival and development.
2. Take 500 μL of the supernatant and add an equal amount of 2× AP buffer, if the AP activity is measured in a spectrophotometer.
3. Do not use buffers containing phosphate, as it is a competitive inhibitor of AP.

4. The entire top of egg can be covered with Sellotape or Scotch tape. This will prevent cracking of the shell during opening a window.
5. Alternatively, DNA injection can be performed using a pressure injection system, such as Picospritzer (Parker).
6. Keep the inside of the incubator clean. Check the eggs in the incubator every day, and discard dead embryos (live embryos have well-developed vasculatures and a clearly visible beating heart).
7. The levels of ectopic expression vary in different embryos. Only the embryos that showed significant levels of expression are to be used for analysis.

References

1. Funahashi J, Okafuji T, Ohuchi H, Noji S, Tanaka H, Nakamura H (1999) Role of Pax-5 in the regulation of a mid-hindbrain organizer's activity. Develop Growth Differ 41 (1):59–72
2. Harada H, Omi M, Nakamura H (2017) In ovo electroporation methods in chick embryos. Methods Mol Biol 1650:167–176. https://doi.org/10.1007/978-1-4939-7216-6_10
3. Hu WY, Myers CP, Kilzer JM, Pfaff SL, Bushman FD (2002) Inhibition of retroviral pathogenesis by RNA interference. Curr Biol 12 (15):1301–1311
4. Katahira T, Nakamura H (2003) Gene silencing in chick embryos with a vector-based small interfering RNA system. Develop Growth Differ 45(4):361–367
5. Harpavat S, Cepko CL (2006) RCAS-RNAi: a loss-of-function method for the developing chick retina. BMC Dev Biol 6:2
6. Koga A, Iida A, Hori H, Shimada A, Shima A (2006) Vertebrate DNA transposon as a natural mutator: the medaka fish Tol2 element contributes to genetic variation without recognizable traces. Mol Biol Evol 23 (7):1414–1419. https://doi.org/10.1093/molbev/msl003
7. Kawakami K, Imanaka K, Itoh M, Taira M (2004) Excision of the Tol2 transposable element of the medaka fish Oryzias latipes in Xenopus laevis and Xenopus tropicalis. Gene 338(1):93–98. https://doi.org/10.1016/j.gene.2004.05.013
8. Kawakami K, Noda T (2004) Transposition of the Tol2 element, an Ac-like element from the Japanese medaka fish Oryzias latipes, in mouse embryonic stem cells. Genetics 166 (2):895–899
9. Kawakami K, Shima A, Kawakami N (2000) Identification of a functional transposase of the Tol2 element, an Ac-like element from the Japanese medaka fish, and its transposition in the zebrafish germ lineage. Proc Natl Acad Sci U S A 97(21):11403–11408. https://doi.org/10.1073/pnas.97.21.11403
10. Kawakami K, Takeda H, Kawakami N, Kobayashi M, Matsuda N, Mishina M (2004) A transposon-mediated gene trap approach identifies developmentally regulated genes in zebrafish. Dev Cell 7(1):133–144. https://doi.org/10.1016/j.devcel.2004.06.005
11. Sato Y, Kasai T, Nakagawa S, Tanabe K, Watanabe T, Kawakami K, Takahashi Y (2007) Stable integration and conditional expression of electroporated transgenes in chicken embryos. Dev Biol 305(2):616–624
12. Nakamoto C, Kuan SL, Findlay AS, Durward E, Ouyang Z, Zakrzewska ED, Endo T, Nakamoto M (2014) Nel positively regulates the genesis of retinal ganglion cells by promoting their differentiation and survival during development. Mol Biol Cell 25 (2):234–244. https://doi.org/10.1091/mbc.E13-08-0453
13. Yee JK, Moores JC, Jolly DJ, Wolff JA, Respess JG, Friedmann T (1987) Gene expression from transcriptionally disabled retroviral vectors. Proc Natl Acad Sci U S A 84(15):5197–5201
14. Invitrogen BLOCK-iT PolII miR RNAi Expression Vector Kits, User Manual. https://www.thermofisher.com/document-connect/document-connect.html?url=https://assets.thermofisher.com/TFS-Assets/LSG/manuals/blockit_miRNAexpressionvector_man.pdf&title=BLOCK-iT™ Pol II miR RNAi Expression Vector Kits

15. Chung KH, Hart CC, Al-Bassam S, Avery A, Taylor J, Patel PD, Vojtek AB, Turner DL (2006) Polycistronic RNA polymerase II expression vectors for RNA interference based on BIC/miR-155. Nucleic Acids Res 34(7): e53. https://doi.org/10.1093/nar/gkl143
16. Flanagan JG, Cheng HJ, Feldheim DA, Hattori M, Lu Q, Vanderhaeghen P (2000) Alkaline phosphatase fusions of ligands or receptors as in situ probes for staining of cells, tissues, and embryos. Methods Enzymol 327:19–35
17. Lee Y, Kim M, Han J, Yeom KH, Lee S, Baek SH, Kim VN (2004) MicroRNA genes are transcribed by RNA polymerase II. EMBO J 23(20):4051–4060. https://doi.org/10.1038/sj.emboj.7600385

Chapter 9

A Simple Guide for Generating BAC Transgenic Animals for Retinal Research

Cavit Agca and Christian Grimm

Abstract

Bacterial artificial chromosomes (BACs) are genomic tools that can carry several hundred kilobases of exogenous genomic material. This allows to incorporate sufficiently large DNA stretches to include most if not all upstream and downstream *cis*-regulatory elements of a gene in order to mimic and analyze its endogenous regulation of expression using a reporter protein in vivo. Here, we illustrate the generation of a BAC:LIF-EGFP transgenic mouse line to describe a simplified version of BAC transgenesis using *galK*-based recombineering.

Key words BAC transgenics, Recombineering, SW102, Electroporation, Colony real-time PCR, *galK*, *Lif*, Minimal media, BAC sequencing

1 Introduction

BAC transgenics is an efficient tool to study endogenous regulation of specific genes in vivo [1]. BAC vectors can carry several hundred kilobases of elements of the gene to study [2, 3]. Once they are transformed into model organisms, modified BACs that are carrying reporter genes facilitate characterization of gene expression with minimal effort.

During the Gene Expression Nervous System Atlas (GENSAT) project, hundreds of BAC transgenic animals that are either expressing Cre recombinase or EGFP instead of the gene of interest were generated in order to characterize expression profiles of genes in mouse neuronal cells [4, 5]. GENSAT transgenic lines are available for research, and approximately 100 of these lines that are expressing EGFP as a reporter were also characterized for retinal expression [6]. Although the GENSAT project covers an important proportion of genes as BAC transgenic reporter lines, there are still many genes that are not available. In addition, a transgenic line may not meet the specific needs for certain projects and cannot be used. Therefore, it often is desirable to generate a new BAC

Chai-An Mao (ed.), *Retinal Development: Methods and Protocols*, Methods in Molecular Biology, vol. 2092, https://doi.org/10.1007/978-1-0716-0175-4_9, © Springer Science+Business Media, LLC, part of Springer Nature 2020 transgenic line that meets the special requirements of specific retinal research projects.

After having decided on the design and confirmed the BAC clone for the gene of interest, it usually takes about 3–4 weeks to modify a BAC clone. The first step is to amplify and transfer the BAC clone from a non-recombineering strain into one of the bacterial recombineering strains [7]. In our case, this is followed by inserting *galK* into the region of interest and finally replacing *galK* by the desired modification in the BAC [8]. To explain the procedures easily, we describe how to engineer a BAC clone that carries a *Lif-EGFP* reporter gene and how to prepare it for the generation of an animal model.

2 Materials

2.1 BAC Preparation for Recombineering

1. BAC clones RP23-451O6 (BAC I) and RP-23-388G19 (BAC II) carrying the mouse *Lif* gene were ordered from BACPAC resources (the RPCI-23 C57BL/6J Mouse BAC Library [9]). Both BACs have the pBACe3.6 backbone which confers chloramphenicol resistance to the bacteria.
2. SW102 strain for *galK* selection (recombineering kit from the National Cancer Institute) [8].
3. Chloramphenicol.
4. PhasePrep BAC DNA isolation kit (Sigma-Aldrich).
5. Restriction enzymes (REs) (NEB): SpeI, XhoI, SmaI, NotI, KpnI, XbaI, and SnaB1.
6. Gel documentation system.
7. Gel electrophoresis and gel casting systems (20–25 cm in length).
8. Incubators at 32 and 37 °C, and shaking water bath at 32, 37, and 42 °C.
9. Electroporation system (Gene Pulser, Bio-Rad).
10. Cuvettes for electroporation (0.1 cm) (Bio-Rad).

2.2 Recombineering

1. *galK and EGFP* expressing plasmids for target amplification.
2. M63 minimal media 5×: 10 g $(NH_4)_2SO_4$, 68 g KH_2PO_4, and 2.5 mg $FeSO_4{\cdot}7H2O$ are dissolved in 1 L ddH_2O. pH is adjusted to 7 with KOH and autoclaved.
3. M9 medium 1×: 7.5 g $Na_2HPO_4{\cdot}2H_2O$, 3 g KH_2PO_4, 1 g NH_4Cl, and 0.5 g NaCl are dissolved in 1 L ddH_2O and autoclaved.
4. Supplements for plates (stock solutions): 0.2 mg/mL d-biotin (sterile filtered), 20% galactose (autoclaved), 20% 2-deoxy-

galactose (DOG) (autoclaved), 20% glycerol (autoclaved), 10 mg/mL L-leucine (heated, cooled, and sterile filtered), and 1 M $MgSO_4 \cdot 7H_2O$.

5. PCR enzyme with proofreading: Expand High-Fidelity PCR System (Roche).
6. DpnI enzyme (Sigma-Aldrich).
7. MacConkey agar.
8. M63 minimal plates + galactose: Autoclave 15 g agar in 800 mL ddH_2O. After initial cooling, add 200 mL of autoclaved 5× M63 minimal media. After medium has been cooled to about 50 °C, add supplements from stock solutions to reach final concentrations of 1 mg/L d-biotin, 0.2% galactose, 45 μg/mL L-leucine, 12.5 μg/mL chloramphenicol, and 1 mM $MgSO_4 \cdot 7H_2O$ and plate (*see* **Note 1**).
9. MacConkey indicator plates + galactose: Autoclave 15 g MacConkey agar in 1 L ddH_2O. After medium has been cooled to about 50 °C, add supplements from stock solutions to reach final concentrations of 0.2% galactose and 12.5 μg/mL chloramphenicol and plate (*see* **Note 1**).
10. M63 minimal plates + glycerol + DOG: Autoclave 15 g agar in 800 mL ddH_2O. After initial cooling, add 200 mL of autoclaved 5× M63 minimal media. After medium has been cooled to about 50 °C, add supplements from stock solutions to reach final concentrations of 1 mg/L d-biotin (sterile filtered), 0.2% glycerol, 0.2% DOG, 45 μg/mL L-leucine, 12.5 μg/mL chloramphenicol, and 1 mM $MgSO_4 \cdot 7H_2O$ and plate (*see* **Note 1**).

2.3 Preparation of BAC for Pronuclear Injection

1. Sepharose CL-4B beads (Sigma-Aldrich).
2. Polyamine supplement (×1000) (Sigma-Aldrich).
3. Microinjection buffer: 0.5 mL 1 M Tris–HCl (pH 7.5), 10 μL 0.5 M EDTA (pH 8.0), and 1 mL 5 M NaCl, 50 μL polyamine supplement (×1000), and complete it to 50 mL ddH_2O. - Sterile filtration and degassing of final solution before use (*see* **Note 2**).

3 Methods

3.1 Deciding on the BAC Clone

1. Check available BAC clones from the BACPAC resources for the gene of interest, preferentially keeping the gene close to the center of the BAC.
2. Try to pick clones that have a minimum number of neighboring genes in order to keep possible gene dosage effects at a minimum [10].
3. If available, order at least two clones since BACs may lose their integrity or have rearrangements due to their big cargo.

3.2 Deciding on the Approach

3.2.1 Reporter Expression In Vivo

1. *Replacing a whole gene with the reporter of interest.* This approach replaces the gene of interest with a reporter, mimicking its endogenous regulation without having a knockout phenotype. However, removal of intronic sequences can interfere with the endogenous regulation. This method also excludes the possibility of overexpressing the target gene.
2. *Insertion of a reporter in frame to the N- or C-terminal without interfering with the gene structure and regulation.* This approach mimics endogenous regulation with the possibility of overexpressing the gene of interest: This method also excludes the possibility of interfering with endogenous regulation by removing intronic sequences (Fig. 1) (*see* **Note 3**).
3. *Partial deletion of a gene while inserting the reporter.* This approach elucidates protein function by removing a specific interaction or a localization domain of the target gene while inserting a reporter. Gene regulatory elements like the 3′UTR region can also be studied by this approach.

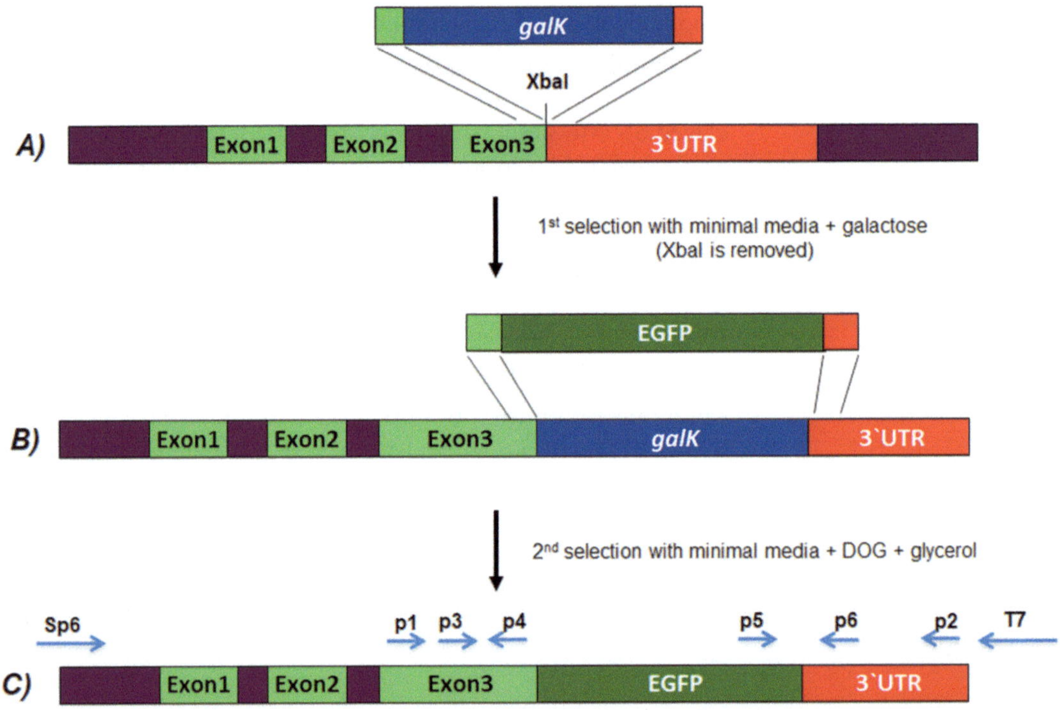

Fig. 1 Recombineering steps and resultant BAC. (**a**) *galK* (blue) target DNA was generated with homology arms of 50 bps identical to *Lif exon 3* (green) and *Lif 3′UTR* (red). Homology arms targeted to the native *Lif* gene at its stop codon and the XbaI cut site. (**b**) *EGFP* target DNA was generated by homology arms of *Lif exon 3* (green) and *Lif 3′UTR* (red). *galK* (blue) is inserted into the *Lif* while removing the stop codon and XbaI site. Recombinant clones (BAC:LIF-galK) were selected by minimal galactose media + galactose by the expression of *galK*. (**c**) BAC:LIF-EGFP is generated by recombineering of BAC:LIF-galK and further selected by a DOG (toxic analog of galactose). Position and direction of primers are shown by blue arrows. SP6 and T7 promoters in pBACe3.6 backbone were used to sequence end points of the final BAC:LIF-EGFP

3.2.2 Overexpression Studies

Pronuclear injection of BAC clones often results in concatemerization of multiple copies of the BAC DNA and their integration into a single locus [11]. Moreover, BAC clones can integrate into multiple loci in the genome. This can result in overexpression of the target transgene as well as the hitchhiker genes within the BAC clone. Multiple insertions to distant loci can be eliminated by backcrossing the transgenic mouse to wild-type animals. Overall, these possibilities give the opportunity to adjust overexpression of the target gene and to study gene dosage effects in model organisms with modified or unmodified BACs.

3.3 BAC Preparation for Recombineering

3.3.1 First Plating and Amplifying the BAC

1. Inoculate ordered BACs (BAC I and BAC II) into 5 mL LB media with chloramphenicol (12.5 μg/mL), and grow overnight at 37 °C with gentle shaking (200 rpm).
2. Plate BAC-containing bacteria to chloramphenicol plates (12.5 μg/mL), and grow overnight in a 37 °C incubator.
3. Pick a single colony, and grow overnight at 37 °C with 5 LB + chloramphenicol (200 rpm).
4. Prepare a glycerol stock for storage.
5. Add 1 mL culture to 500 mL (*see* **Note 4**) LB + chloramphenicol for BAC DNA isolation.
6. Isolate BACs using PhasePrep BAC DNA kit from Sigma-Aldrich according to manufacturer's instructions (*see* **Note 5**).
7. Dissolve the BAC DNA pellet overnight at 4 °C, and do not freeze the BAC DNA afterwards.
8. Take OD_{260}/OD_{280} measurements.

3.3.2 Ensuring that the BAC Is Correct: RE Digestion and PCR

1. Identify restriction enzymes that cut BACs in 20–30 fragments using any restriction digestion software. SpeI usually gives an ideal pattern (Fig. 2a). For BAC I and BAC II, we used the BAC sequences obtained from BACPAC resources. Additional enzymes were also used for analysis: XhoI, SmaI, NotI, KpnI, and SnaB1.
2. Digest different concentrations of BACs (0.5, 1, and 1.5 μg) with 1 unit of SpeI enzyme overnight at 37 °C in a final volume of 20 μL (Fig. 2b).
3. For BAC I and BAC II, digestion with XhoI, SmaI, NotI, KpnI, and SnaB1 enzymes is also done.
4. Run digested BACs in 0.7% agarose at 30 V overnight using at least a 20–25 cm gel electrophoresis and gel casting system.
5. Take images of gels (*see* **Notes 6** and 7).
6. Cross compare the digestion patterns of BACs to the estimated pattern obtained by the software (*see* **Note 8**) (Fig. 2).

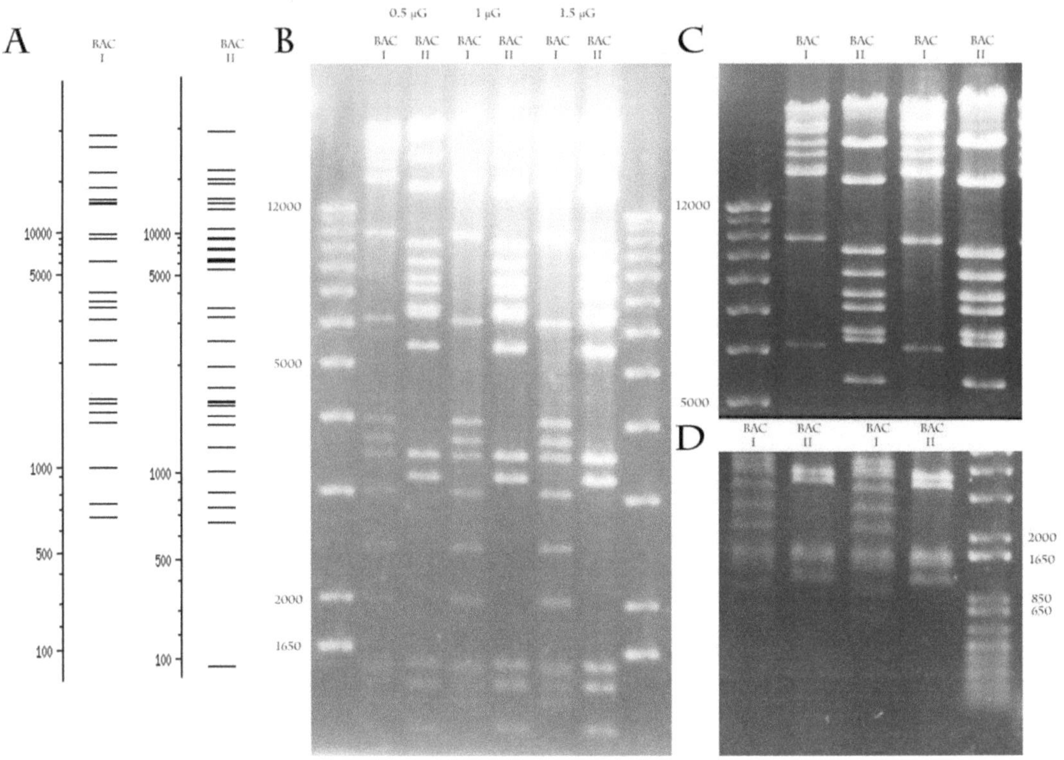

Fig. 2 SpeI digestion of Lif BACs. (**a**) Predicted patterns of SpeI digestion of BAC I and BAC II. (**b**) Digestion of BAC I and BAC II at different concentrations. Gel images focusing at high molecular weight bands (**c**) and lower molecular weights (**d**)

7. BAC was further confirmed by PCR amplification of the region between exon 3 and whole 3′UTR of *Lif* (Fig 1a).

   PCR set up: 100 ng BAC DNA, 0.75 μL 2.6 unit Roche high-fidelity enzyme, 5 μL 10× Roche high-fidelity enzyme buffer with 15 mM $MgCl_2$, 1 μL dNTP mix (200 μM each), and primers p1 and p2 (Fig. 1c, Table 1) (300 nM, final concentration for each), and nuclease-free H_2O was added to generate 50 μL total reaction volume.

   PCR conditions for fragment lengths above 3 kb:

   Initial denaturation: 94 °C 2 min

   Amplification: 94 °C, 15 s/60 °C, 30 s/68 °C, 3 min; 30 cycles

   Final extension: 68 °C, 7 min

8. Digest 10 μL of PCR product with 1 unit of the enzyme of interest for 1 h at 37 °C in a final volume of 20 μL. XbaI was used for BAC I and BAC II as it cuts *Lif* at its stop codon (Fig. 1a).
9. Digested PCR products are run on a 2% agarose gel, and images of the gel are analyzed.

Table 1
Primer sequences that were used for recombineering and analysis of resultant BAC

BAC transgenesis primers	
LIF-GALK F	AGCTTCTGGGGACATACAAGCAAGTCATAAGTGTGGTGGTCCAGGCC TTCCCTGTTGACAATTAATCATCGGCA
LIF-GALK R	GATCCTGCTCCTGAGGTCCCTCAGTCCATGGTACATTCAAGACCTCCTC TTCAGCACTGTCCTGCTCCTT
LIF-EGFP F	AGCTTCTGGGGACATACAAGCAAGTCATAAGTGTGGTGGTCCAGGCCTTCA TGGTGAGCAAGGGCGAGGAGC
LIF-EGFP R	GATCCTGCTCCTGAGGTCCCTCAGTCCATGGTACATTCAAGACCTCCTC TTTACTTGTACAGCTCGTCCAT
pl	GTGCGCCTAACATGACAGACTTC
p2	GAGTTCTTTATTATTTCATTTTT
P3	CCATGGCAACGGGACAGAGA
p4	ACGGCAGTGGGGTTCAGGAC
p5	CGCCGCCGGGATCACTCTCG
p6	GTCCCAAACCCCAGCACATT

3.4 Transformation of BAC I into Recombinogenic Bacterial Strain SW102

3.4.1 Preparation of Electro-Competent SW102

1. Prepare an overnight culture of SW102 in 5 mL LB using SW102 glycerol stock.
2. Transfer 500 μL into 25 mL LB + chloramphenicol.
3. Incubate for 3–5 h at 32 °C in a shaking water bath until the density reaches an OD_{600} of 0.6.
4. Optional step: Transfer 10 mL to another 50 mL flask for heat shock. Heat shock can be performed at this stage by incubation at 42 °C for exactly 15 min in a shaking water bath to activate recombineering (*see* **Note 9**).
5. Put the media with the bacteria on ice/water, and transfer 10 mL into precooled 15 mL Falcon tubes.
6. Spin down at 0 °C for 5 min (5000 × *g*).
7. Pour off all supernatant, and briefly invert tube on a paper towel to drain remaining liquid.
8. Resuspend the pellet in 1 mL ice-cold ddH_2O by gently shaking the tube in the ice/water, and fill up to 10 mL with ice-cold ddH_2O.
9. Repeat **steps 6** and 7.
10. Resuspend electro-competent cells in approximately 50 μL ice-cold ddH_2O.

3.4.2 Electroporation of BAC into SW102

1. Mix 25 μL of electro-competent cells with 2 μg of BAC DNA in a precooled Eppendorf tube.
2. Transfer the mix into a precooled electroporation cuvette (0.1 cm).
3. Electroporate using the condition: 1.75 kV, 25 μF, and 200 Ω.
4. Add 1 mL LB media, and incubate for 1 hour in a shaking water bath at 32 °C.
5. Plate different amounts of electroporated cells on chloramphenicol plates (e.g., 100 μL, 100 μL of a 1:10 dilution, and 100 μL of a 1:100 dilution), and incubate at 32 °C overnight.

3.4.3 Confirmation of the BAC Clone in SW102 (SW102: Lif)

1. Select single colonies, and incubate in 5 mL LB + chloramphenicol.
2. Initial confirmation of BAC I was done by colony real-time PCR using primers p3 and p4 (Fig. 1c, Table 1) for the presence of *Lif* genomic DNA.

   Real-time PCR reaction: 1 μL SW102:*Lif* in LB media, 5 μL SYBR Green Master Mix (Applied Biosystems), 1.2 μL primer mix of p3 and p4 (Fig. 1c, Table 1) (2.5 μM each), and 2.8 μL nuclease-free H_2O for 10 μL total reaction volume.
3. BAC I was isolated from positive SW102:*Lif* colonies (*see* Subheading 3.3.1, **steps 4–8**) and analyzed by RE digestion (*see* Subheading 3.3.2, **steps 2–6**).
4. Further analysis of BAC I was done by PCR and RE digestion with primers p1 and p2 (Fig. 1c, Table 1) (*see* Subheading 3.3.2, **steps 7–9**).

3.5 Recombineering

3.5.1 galK Insertion by Recombineering (SW102: Lif-galK)

1. *galK* target DNA was prepared by PCR reaction using DNA polymerase with proofreading activity.
2. Each *galK* primer was designed with an addition of 50 bp homology arm against *Lif* genomic sequences (Fig. 1a and Table 1) (*see* **Note 10**).
3. PCR was done with 5 ng *galK* plasmid DNA and *galK* primers with homology arms using the conditions below (*see* Subheading 3.3.2, **step 7**).

   PCR conditions for fragment lengths below 3 kb:

   Initial denaturation: 94 °C, 2 min

   Amplification: 94 °C, 15 s/60 °C, 30 s/72 °C, 2 min; 30 cycles

   Final extension: 72 °C, 7 min
4. PCR product was digested on spot by adding 20 units of DpnI (digests only methylated DNA) for 1 h at 37 °C to remove template plasmid.
5. PCR product was column purified and OD_{260}/OD_{280} was measured.

3.5.2 Electroporation of galK Target DNA into SW102:Lif

1. Electro-competent bacteria was prepared from SW102:*Lif* strain with the heat shock step (*see* Subheading 3.4.1, **steps 1–10**). Two different electro-competent cells were generated: heat shock induced for recombineering and uninduced as a control.
2. Mix 25 μL of electro-competent cells with 80–100 ng of *galK* target DNA in a precooled Eppendorf tube for both induced and uninduced cells, separately (*see* **Note 11**).
3. Transfer the mix into a precooled electroporation cuvette (0.1 cm).
4. Electroporate using the condition: 1.75 kV, 25 μF, and 200 Ω.
5. Add 1 mL LB media, and incubate 1 h in a shaking water bath at 32 °C.
6. Centrifuge 1 min at 16,000 × *g*, and wash with M9 media (*see* **Note 12**).
7. Repeat centrifugation step, wash with M9 media, and resuspend pellet.
8. Plate different amounts of the resuspended bacterial pellet using M63 minimal plates + galactose (e.g., 100 μL, 100 μL of a 1:10 dilution, and 100 μL of a 1:100 dilution) both for induced cells and uninduced controls (*see* **Note 13**).
9. Incubate at 32 °C for 3 days.

3.5.3 Selection and Confirming galK Integration

1. Initial confirmation of *galK* insertion was done by colony PCR. Clones from M63 minimal plates + galactose plate were picked using a sterile tip (label the colonies at the bottom of the plate) and used for colony PCR using primers p1 and p2 (Fig. 1c, Table 1) which was followed by XbaI digestion (*see* Subheading 3.3.2, **steps 7–9**). XbaI site should have been removed from correctly recombineered colonies (Fig. 1b).
2. Streak a few positive colonies that were previously labeled onto MacConkey indicator plates + galactose, and grow overnight at 32 °C. This is a critical step to exclude hitchhiker strains (*see* **Note 14**).
3. Select bright-red colonies, and grow in 5 mL LB + chloramphenicol.
4. Isolate BAC (*see* Subheading 3.3.1, **steps 4–8**), and analyze by RE digestion (*see* Subheading 3.3.2, **steps 2–6**)

3.5.4 EGFP Insertion by Recombineering

1. *galK* recombineering procedure was followed using *EGFP* primers with 50 bp homology arms identical to the *Lif* homology arms in *galK* target DNA, *EGFP* plasmid, and SW102:*Lif-galK* from the previous step (*see* Subheading 3.5.1, **steps 1–5**, and Subheading 3.5.2, **steps 1–7**) (Fig. 1b and Table 1).

2. M63 selection plates with DOG and glycerol supplement was used for selection (*see* **Note 15**). Different amounts (e.g., 100 μL, 100 μL of a 1:10 dilution, and 100 μL of a 1:100 dilution) were plated both for induced cells and uninduced controls.
3. Incubate at 32 °C for 3 days.

3.5.5 Selection and Confirming EGFP Integration

1. Initial confirmation of BACs was done by colony real-time PCR using primers p5 and p6 (Fig. 1c, Table 1) (*see* Subheading 3.4.3, **steps 1–2**).
2. Isolate a positive BAC (*see* Subheading 3.3.1, **steps 4–8**), and analyze by RE digestion (*see* Subheading 3.3.2, **steps 1–6**). An expected band shift was observed at recombination site after KpnI digestion of BAC:LIF-EGFP (Fig. 3a).
3. Sequence BAC after PCR amplification of the modified region (*see* Subheading 3.3.2, **step 7**) as well as direct sequencing of the BAC which is offered by several companies.

3.6 Preparation of BAC for Injection and Analysis of F1 Generation

3.6.1 Preparation of BAC for Injection

1. SW102:*Lif-EGFP* was grown overnight at 37 °C with 5 LB + chloramphenicol, and BAC:LIF-EGFP was isolated (*see* Subheading 3.3.1, **steps 4–8**).
2. 50 μg of BAC:LIF-EGFP was digested with 100 unit NotI enzyme overnight at 37 °C in a final volume of 300 μL. NotI restriction site is not present within BAC:LIF-EGFP, and it completely removes the pBAC3e.6 backbone.
3. NotI was heat inactivated at 65 °C for 15 min.
4. Set up a sepharose column using a 5 mL serological pipette. Move the cotton plug of the pipette to the tip [10].
5. CL-4B sepharose beads are loaded into column/pipette using a syringe and tubing that fits to the pipette. It is important to avoid generating bubbles.
6. 200 μL of microinjection buffer and 10 μL of 6× DNA loading dye were mixed with the digest.
7. Digested BAC was run in sepharose column.
8. When the liquid level reached the surface of the sepharose beads, 1 mL injection buffer was added, and collection of 500 μL fractions was started.
9. Repeat adding 1 mL injection buffer until 10 mL has been reached, and keep collecting fractions.
10. Run a gel and check each fraction. Fraction 8 gave best results (Fig. 3b).
11. Take OD_{260}/OD_{280} measurements.
12. Fraction 8 (8 ng/μL) was diluted to 1 ng/μL with microinjection buffer to get ready for pronuclear injections.

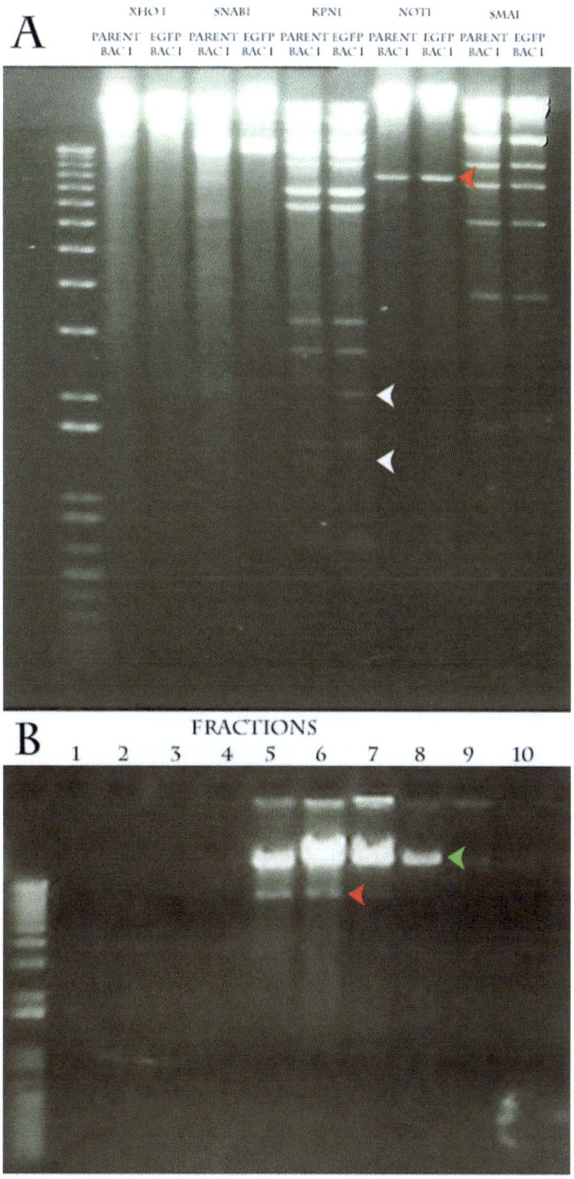

Fig. 3 Confirmation and preparation of BAC:LIF-EGFP. (**a**) BAC:LIF-EGFP and the parent BAC I were digested with XhoI, SmaI, NotI, KpnI, and SnaB1 enzymes for comparison. KpnI digestion clearly showed an expected 700 bp shift due to *EGFP* integration (white arrows). NotI digestion removed pBACe3.6 backbone (red arrow). (**b**) Column-purified BAC:LIF-EGFP after NotI digestion. pBACe3.6 backbone after digestion is shown by red arrow. Green arrow shows purified BAC in fraction 8 which was purified from the backbone and other genomic leftovers. This fraction is used for pronuclear injections

3.6.2 Selection and Confirmation of BAC: LIF-EGFP Transgenic Animals

1. Approximately 50 pups were obtained from the pronuclear injections.
2. Genomic DNA from biopsies of the newborn mice was isolated, and PCR reaction was performed to detect BAC transgenic animals (*see* Subheading 3.3.2, **step** 7) (*see* **Note 16**).
3. Further confirmation was done by sequencing of modified regions after PCR amplification.

4 Notes

1. All selection plates have to be prepared freshly due to possible degradation of supplements.
2. Microinjection buffer has to be prepared fresh for each usage.
3. Chimeric gene product with the reporter can be functional.
4. 500 mL BAC culture approximately yields around 100 μg of BAC DNA.
5. Use orifice tips during isolation procedure to minimize damage to BAC integrity.
6. As the signal intensity and separation dynamics depends on the size of the bands, different sizes of the bands may need isolated imaging (Fig. 2c, d).
7. It is important to take images at early time points for band sizes of 1 kb or less as the bands fade during an overnight gel run (Fig. 2d).
8. BAC II had several bands missing and was therefore excluded from further analysis.
9. No heat shock at 42 °C is necessary at this step. Heat shock is for induction of recombinant protein expression.
10. Primers should be column purified.
11. Ideally, 10–30 ng target DNA should be enough for electroporation, but different amounts of target DNA can be used at this point (up to 200 ng) to optimize the colony yield. In our hands, 80–100 ng target DNA gave the optimal number of colonies.
12. Washing with M9 media is critical for plate selection due to possible carryover of carbon sources.
13. SW102 strain has a gal operon with the deletion of *galK*. Therefore, introduction of *galK* to SW102 that is grown in minimal medium containing galactose as only carbon source will be a positive selection for recombineering.

14. MacConkey + galactose indicator plates are important to get rid of Gal− strains (hitchhikers) which are detected by white colonies. Gal+ bacteria become bright red/pink because of the pH change due to fermented galactose.
15. Gal+ bacteria will be eliminated due to DOG (toxic galactose analog) negative selection. Gal− bacteria will only use glycerol as carbon source and will be positively selected for *galK* removal after recombineering.
16. Do not exclude animals that have weak positive bands. This may refer to a chimeric animal that may become fully positive at F2 generation.

Acknowledgments

We would like to thank Dr. Magdalena Anna Krzyzaniak (Institute of Biochemistry, ETH Zurich) for sharing reagents for recombineering and Dr. Pawel Pelczar (Institute for Laboratory Animal Science, University of Zurich) for pronuclear injections.

References

1. Yang XW, Model P, Heintz N (1997) Homologous recombination based modification in Escherichia coli and germline transmission in transgenic mice of a bacterial artificial chromosome. Nat Biotechnol 15(9):859–865. https://doi.org/10.1038/nbt0997-859
2. O'Connor M, Peifer M, Bender W (1989) Construction of large DNA segments in Escherichia coli. Science 244 (4910):1307–1312
3. Shizuya H, Birren B, Kim UJ, Mancino V, Slepak T, Tachiiri Y, Simon M (1992) Cloning and stable maintenance of 300-kilobase-pair fragments of human DNA in Escherichia coli using an F-factor-based vector. Proc Natl Acad Sci U S A 89(18):8794–8797
4. Gong S, Zheng C, Doughty ML, Losos K, Didkovsky N, Schambra UB, Nowak NJ, Joyner A, Leblanc G, Hatten ME, Heintz N (2003) A gene expression atlas of the central nervous system based on bacterial artificial chromosomes. Nature 425(6961):917–925. https://doi.org/10.1038/nature02033
5. Schmidt EF, Kus L, Gong S, Heintz N (2013) BAC transgenic mice and the GENSAT database of engineered mouse strains. Cold Spring Harb Protoc 2013(3). https://doi.org/10.1101/pdb.top073692
6. Siegert S, Scherf BG, Del Punta K, Didkovsky N, Heintz N, Roska B (2009) Genetic address book for retinal cell types. Nat Neurosci 12(9):1197–1204. https://doi.org/10.1038/nn.2370
7. Lee EC, Yu D, Martinez de Velasco J, Tessarollo L, Swing DA, Court DL, Jenkins NA, Copeland NG (2001) A highly efficient Escherichia coli-based chromosome engineering system adapted for recombinogenic targeting and subcloning of BAC DNA. Genomics 73(1):56–65. https://doi.org/10.1006/geno.2000.6451
8. Warming S, Costantino N, Court DL, Jenkins NA, Copeland NG (2005) Simple and highly efficient BAC recombineering using galK selection. Nucleic Acids Res 33(4):e36. https://doi.org/10.1093/nar/gni035
9. Osoegawa K, Tateno M, Woon PY, Frengen E, Mammoser AG, Catanese JJ, Hayashizaki Y, de Jong PJ (2000) Bacterial artificial chromosome libraries for mouse sequencing and functional analysis. Genome Res 10(1):116–128
10. Johansson T, Broll I, Frenz T, Hemmers S, Becher B, Zeilhofer HU, Buch T (2010) Building a zoo of mice for genetic analyses: a comprehensive protocol for the rapid generation of

BAC transgenic mice. Genesis 48(4):264–280. https://doi.org/10.1002/dvg.20612

11. Van Keuren ML, Gavrilina GB, Filipiak WE, Zeidler MG, Saunders TL (2009) Generating transgenic mice from bacterial artificial chromosomes: transgenesis efficiency, integration and expression outcomes. Transgenic Res 18(5):769–785. https://doi.org/10.1007/s11248-009-9271-2

Chapter 10

Identification and Characterization of Cis-Regulatory Elements for Photoreceptor-Type-Specific Transcription in ZebraFish

Wei Fang, Yi Wen, and Xiangyun Wei

Abstract

Tissue-specific or cell-type-specific transcription of protein-coding genes is controlled by both *t*rans-*r*egulatory *e*lements (TREs) and *c*is-*r*egulatory *e*lements (CREs). However, it is challenging to identify TREs and CREs, which are unknown for most genes. Here, we describe a protocol for identifying two types of transcription-activating CREs—core promoters and enhancers—of zebrafish photoreceptor type-specific genes. This protocol is composed of three phases: bioinformatic prediction, experimental validation, and characterization of the CREs. To better illustrate the principles and logic of this protocol, we exemplify it with the discovery of the core promoter and enhancer of the *mpp5b* apical polarity gene (also known as *ponli*), whose red, green, and blue (RGB) cone-specific transcription requires its enhancer, a member of the rainbow enhancer family. While exemplified with an RGB-cone-specific gene, this protocol is general and can be used to identify the core promoters and enhancers of other protein-coding genes.

Key words Gene expression regulation, Transcription factors, CREs, Cis-regulatory elements, Bioinformatics, Zebrafish, Teleost, Photoreceptor, Retina, Apical polarity genes, *ponli*, *nagie oko*, *mpp5a*, *mpp5b*

1 Introduction

In eukaryotes, tissue-specific or cell-type-specific transcription of each protein-coding gene is regulated by specific *t*rans-*r*egulatory *e*lements (TREs) and *c*is-*r*egulatory *e*lements (CREs). Both TREs and CREs can be divided into multiple categories. TREs encompass RNA polymerase II, transcription factors, transcriptional coregulators, chromatin modifiers, regulatory RNAs, etc. [1–3]. And CREs encompass core promoters, enhancers, locus control regions, silencers, and insulators [4–7]. TREs and CREs regulate transcription through complex interactions. At the center of these interactions are the specific bindings between transcription factors and CREs [4, 7]. Thus, identifying CREs is an essential step toward understanding the spatiotemporal transcription of a gene.

Chai-An Mao (ed.), *Retinal Development: Methods and Protocols*, Methods in Molecular Biology, vol. 2092,
https://doi.org/10.1007/978-1-0716-0175-4_10,

Each category of CREs plays different roles in transcription. Core promoters mediate the docking of the transcription pre-initiation complex by recruiting basic transcription factors; thus, they are required for transcription initiation. Core promoters are about 100–200 bp long and contain transcription start sites (TSSs) [8]. By contrast, enhancers, spanning about 80–250 bp, recruit additional transcription factors to activate core-promoter-bound transcription initiation complex; thus, enhancers are required for transcription completion, often activating transcription in a spatiotemporally specific manner [9, 10]. Unlike core promoters, enhancers locate away from TSSs, often thousands of base pairs away. However, in some genes, transcriptional specificity-deciding cis-elements locate very close to the TSSs, around 100–200 bp in distance; these elements are sometimes referred to as promoter-proximal elements or proximal promoters in the literature [8, 11]. Despite their proximity to TSSs, promoter-proximal elements or proximal promoters are functionally like enhancers [11]. For the sake of simplicity and clarity, here, we encompass promoter-proximal elements or proximal promoters in the category of enhancers per their function of stimulating transcription. Thus, it needs to be kept in mind that the compositions and locations of enhancers can vary drastically from gene to gene.

Like enhancers, locus control regions activate specific transcription, but they exert broader effects by controlling a cluster of related genes within a locus [12]. In contrast to enhancers, silencers suppress transcription by recruiting transcription suppressors [6, 13]. Unlike enhancers, locus control regions, and silencers, all of which act directly on their target genes, insulators act indirectly by preventing enhancers, silencers, and locus control regions on one side of an insulator from affecting genes on the opposite side of the insulator [14]. These CRE properties suggest that core promoters and enhancers are the basic transcription-activating elements that mediate spatiotemporal transcription of a gene, making these two CREs useful for practical applications, such as expressing therapeutic genes in specific cell types in gene therapies. Despite their importance, the core promoters and enhancers of most genes remain unknown and challenging to identify, particularly enhancers.

Here, we describe a protocol for identifying the core promoters and enhancers of zebrafish photoreceptor-type-specific genes. This protocol is composed of three phases: Phase I predicts the CREs with bioinformatic algorithms, Phase II validates the candidate CREs in vivo with transgenic expression assays, and Phase III characterizes the function of CREs and their motifs by element-swapping and sequence motif-mutating analyses. To better illustrate the principles and the logic of this protocol, we exemplify the protocol with the discovery of the core promoter and enhancer of the *mpp5b* zebrafish apical polarity gene (also known as *ponli*)

[15]. *mpp5b* is restrictively expressed in red, green, and blue (RGB) cones, and this specific transcription needs its enhancer, which is a member of the rainbow enhancer family [16]. Although exemplified with an RGB cone-specific gene, this protocol has broad applications and can be used to identify the core promoters and enhancers of other tissue-specific or cell type-specific protein-coding genes.

2 Materials

2.1 Fish Lines and Fish Care

Zebrafish strains can be purchased from the Zebrafish International Resource Center (https://zebrafish.org/home/guide.php). Medaka fish can be obtained from the Arizona Aquatic Garden (https://www.azgardens.com/). Zebrafish and medaka can be maintained on a 14-h light/10-h dark cycle. Please follow institutional regulations on experimental animal care.

2.2 Bioinformatics Tools

1. The UCSC Genome Browser (http://genome.ucsc.edu/) can be used to browse, extract, and compare genomic sequences of many model species.
2. The MEME Suite (http://meme-suite.org/), a bioinformatics suite of motif-based sequence analysis tools, can be used to identify short nucleotide or peptide sequence motifs and to search for them in long DNA or amino acid sequences.
3. TRANSFAC Professional (http://gene-regulation.com/), a web-based database of the known transcription factors, can be used to predict binding sites for known transcription factors in DNA sequences.

2.3 Reagents for Recombinant DNA Technologies

1. The Mini-Genomic DNA Buffer Set (QIAGEN, Cat #: 19060) can be used to isolate the genomic DNAs from animal tissue samples.
2. The Platinum PCR SuperMix High-Fidelity DNA polymerase (Invitrogen, Cat #: 12532016) can be used to amplify DNA fragments with a high degree of accuracy.
3. The QIAGEN Plasmid Mega Kit (Cat #: 12181) can be used to isolate DNA from BAC or PAC clones.
4. The Q5 Site-Directed Mutagenesis Kit (NEB, Cat #: E0554S) can be used to make deletion and substitution mutations of DNA constructs.
5. I-SceI, a homing endonuclease that recognizes and cuts the target sequence of TAGGGATAACAGGGTAAT (NEB, Cat #: R0694S), is recommended to be coinjected with the pSceI-based transgenic constructs to generate transgenic zebrafish [17].

3 Equipment

1. A Flaming/Brown Micropipette Puller (Model P-97; Sutter Instrument Co.) can be used to prepare embryo injection needles with borosilicate glass capillaries (Kwik-FilTM; World Precision Instruments, Inc. Cat #:1B100-4) [18, 19].
2. A Pneumatic PicoPump (World Precision Instruments, Inc.; PV820) can be used to inject DNA constructs into zebrafish embryos [18, 19].

4 Methods

4.1 Phase I: Bioinformatic Prediction of the Core Promoter and Enhancer of a ZebraFish Gene

4.1.1 Rationale

Core promoters and enhancers can be predicted with many bioinformatic algorithms. Such prediction is based on the logic that the DNA sequences that possess characteristic properties of CREs are likely CREs. The following three characteristic properties are particularly important for predicting core promoters and enhancers: locations, sequence conservation, and possession of transcription factor binding sites.

The locations of core promoters and enhancers: Core promoters contain TSSs [4, 8, 9, 20]. By contrast, enhancers do not overlap with TSSs and often reside thousands of base pairs away from TSSs, up to 100 kb in distance. (In some genes, enhancers may also reside not far from TSSs, like 100–200 bp away from TSSs.) Enhancers can localize either upstream or downstream of the TSSs [10]. When residing downstream of a TSS, an enhancer can either locate within an intron or downstream of the gene. This versatility of enhancer locations makes them more challenging to identify than core promoters.

Sequence conservation of core promoters and enhancers: Core promoters and enhancers generally reside in conserved noncoding regions, thus flanked by un-conserved DNA sequences. Such sequence conservations exist both among orthologs of multiple species and among coregulated different genes within a single genome [21–24].

Possession of binding sites for transcription factors: Core promoters and enhancers are densely packed with highly conserved short sequence motifs (between 6 and 17 bp), some of which are palindromic or direct repeats. These sequence motifs are binding sites for transcription factors [4, 25], which often bind to DNA as dimers or even trimers, with each transcription factor binding to a sequence as short as 3 nucleotide residues [1, 3, 26]. For example, an AT-rich TATA box, normally residing 30 bp upstream of the TSSs of some core promoters, recruits TBP of the TFIID complex [27], and an initiator, spanning the TSSs, recruits a TAFII250–TAFII150 dimer [28]. Similarly, multiple binding sites in enhancers recruit various transcription factors which cooperate to mediate tissue-specific or cell type-specific transcription [26].

The abovementioned properties can be used to predict core promoters and enhancers with bioinformatic tools via one or both of two strategies: The first strategy utilizes sequence conservation among the CREs of coregulated but different genes in a single genome [29–33]. While powerful, this strategy requires prior knowledge of co-transcription profiling, which might be missing or incomplete for the genes of interest. In addition, this strategy works less successfully in higher organisms than in yeast and other lower organisms, presumably because the CREs in higher organisms are often not restricted to nearby upstream sequences and because their motif organizations are more complex.

The second strategy, namely, the phylogenetic footprinting strategy [34], utilizes sequence conservation among the CREs of the orthologous genes of multiple species [35]. Although independent of prior knowledge of co-transcription profiling as required for the first strategy, the phylogenetic footprinting strategy can also run into problems when the orthologous sequences are either too closely related, which makes a global multiple alignment uninformative, or too distantly related, which could make it impossible to align conserved motifs [35]. However, with the genomic sequences of more species becoming available, the phylogenetic footprinting strategy becomes more and more practical.

Thus, Phase I of this protocol takes the phylogenetic footprinting strategy to predict the core promoter and enhancer of a zebrafish gene with web-based algorithms (Fig. 1). In the following, we exemplify the procedures of such prediction with the *mpp5b* gene.

4.1.2 Procedures

1. Identify teleost orthologs of a zebrafish gene: Several teleost orthologs of a zebrafish gene can be identified at the UCSC Genome Browser with the BLAT algorithm, which searches for genomic regions that are homologous to query sequences (http://genome.ucsc.edu/; [36]; for an introduction to the UCSC Genome Browser [37]; for video tutorials on the UCSC Genome Browser, visit http://genome.ucsc.edu/training/index.html).

   For example, to locate the teleost orthologs of the zebrafish *mpp5b* gene, open the BLAT search page at http://genome.ucsc.edu/cgi-bin/hgBlat, which can also be reached by clicking on the "BLAT" button at the home page of the UCSC Genome Browser. Then select "Medaka" from the genome options, and leave other parameters in their default settings. (Searching the medaka genome works better for *mpp5b* than searching other fish genomes because the orthologs of more teleost species were aligned, thus maximally revealing sequence conservations. For other genes, we recommend trying all available non-zebrafish teleost genomes for the

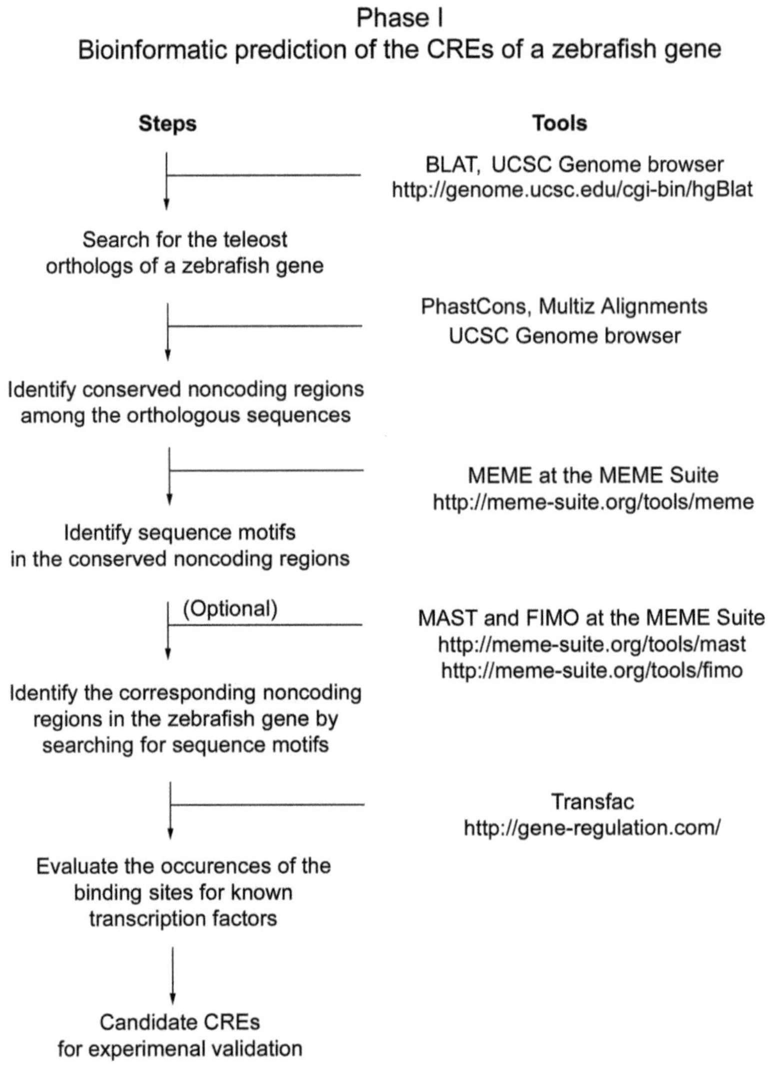

Fig. 1 The outline of Phase I of the protocol—CRE prediction. The flowchart shows the steps to predict CREs with some web-based bioinformatic tools

best outcomes. Please note that teleost genomes used for evolutionary conservation analyses might have not been updated.) Next, paste the entire amino acid sequence of the zebrafish Mpp5b protein (gene accession number GU197553) in the query sequence box (please note that amino acid sequences serve better than nucleotide sequences as queries for cross-species BLAT searches), and then click on the "submit" button to reveal two medaka genes to be identified, one on chromosome 24 with higher matching scores than the other, on chromosome 20 (Fig. 2a). Then click on the "browser" button of the gene with higher scores (i.e., the medaka ortholog of zebrafish *mpp5b*) to display the medaka genomic region,

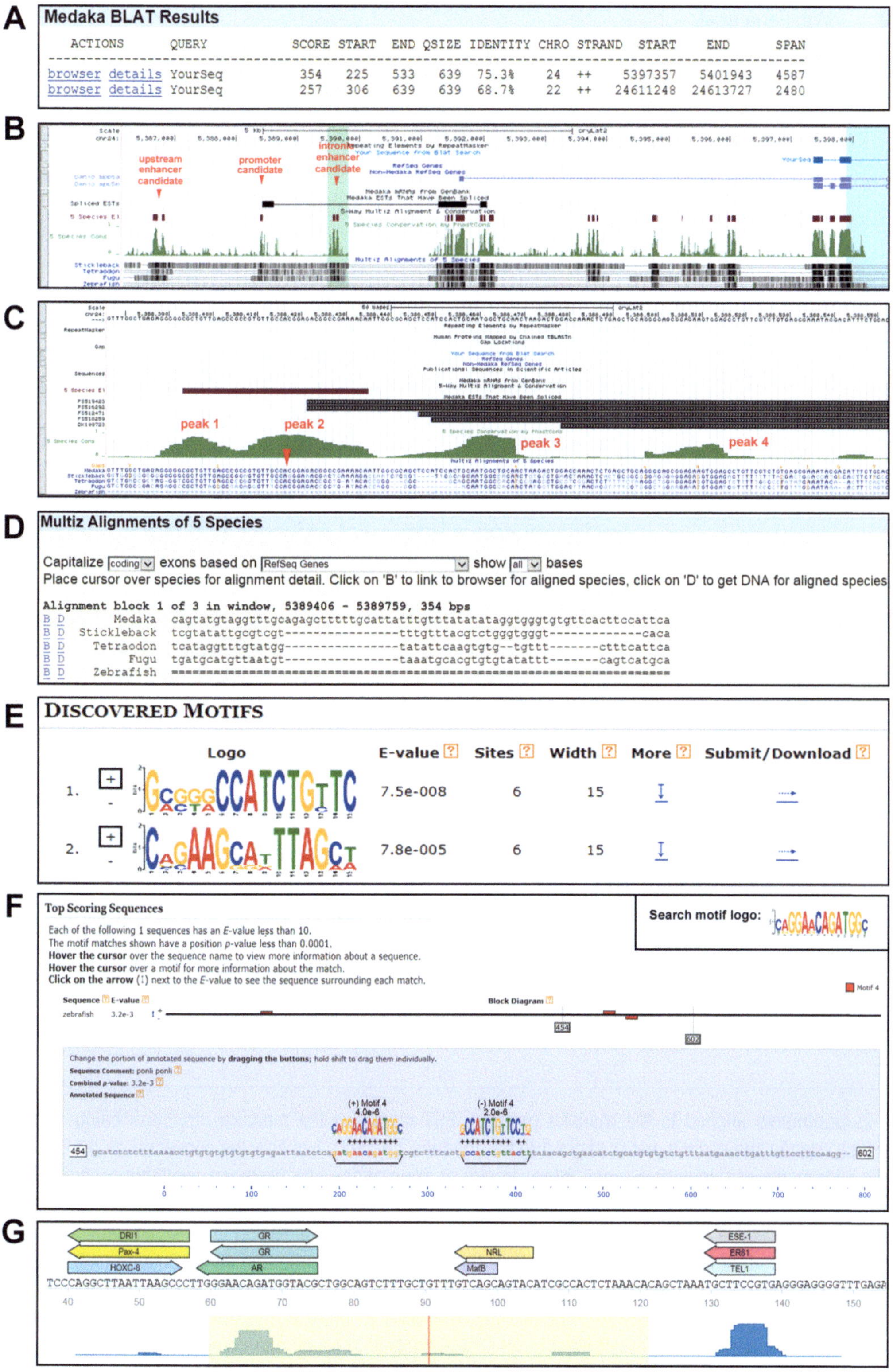

Fig. 2 Example results of the bioinformatic analyses. (**a**) A result page of a BLAT search of the medaka genome with the amino acid sequence of zebrafish Mpp5b protein as the query, showing two medaka homologs identified. (**b**) A user interface page at the UCSC Genome Browser displays the query of zebrafish Mpp5b

which the *mpp5b* gene matches (Fig. 2b). (Please note that the amino acid query sequence of zebrafish Mpp5b is split into sections and matched only to the coding regions of the predicted medaka *mpp5b* gene. The noncoding sequences of the medaka *mpp5b* mRNA are not marked because medaka *mpp5b* has yet to be verified experimentally and to be annotated as such.)

2. Identify the TSS of the gene of interest: It is important to identify TSSs because they are the dividing points between the upstream and downstream sequences of genes. However, TSSs need to be extrapolated from the 5′ ends of mRNA sequences, which can be obtained experimentally by 5′ RACE analyses (the 5′ RACE System for Rapid Amplification of cDNA Ends, version 2.0; Thermo Fisher Scientific, Cat #: 18374058). Alternatively, one can estimate TSSs from the 5′ end sequences of expressed sequence tags (ESTs). Please note that EST sequences may not extend to TSSs; thus, they may not pinpoint TSSs. Please also note that even if ESTs actually cover TSSs, different ESTs may suggest different TSSs because the core promoter of a gene can be a dispersed type of core promoter which possesses multiple close-spaced TSSs [8, 38].

   For example, to estimate the TSS of the medaka *mpp5b* gene, whose full-length cDNA sequence has not been published, we can utilize its EST sequences. Two sets of EST sequences are available, one set for the 5′ end of the gene and the other for its 3′ end. Drag the interactive screen of the UCSC Genome Browser to the upstream direction to reveal the five 5′ end ESTs (FS519423, FS516292, FS12471, FS518259, and DK109723) (Fig. 2c). Among them, FS519423 and FS516292 extend farthest in the upstream direction, making their 5′ ends the best estimate of the TSS, although the TSS of zebrafish *mpp5b* [15] and the sequence

Fig. 2 (continued) aligned to the medaka genome, EST clones of the medaka *mpp5b* ortholog, published zebrafish *mpp5a* and *mpp5b* genes aligned to the medaka genome, five species conservation by PhastCons, Multiz alignments of five species, etc. (More tracks of annotations can be made available by adjusting the track settings.) Red arrowheads indicate the locations of some core promoter and enhancer candidates of the *mpp5b* orthologs. (**c**) A page at the UCSC Genome Browser displays the region around the transcription start site (red arrowhead) of the *mpp5b* orthologs, showing the four conservation peaks by PhastCons (peaks 1–4). Hyphens, sequence gaps; equal signs, un-alignable regions. (**d**) A page at the UCSC Genome Browser displays the Multiz alignments of the *mpp5b* orthologs and the links to browse and download the sequences of a selected genomic region. (**e**) A result page of an MEME analysis displays the consensus sequences of two of many motifs in the enhancers of *mpp5b* genes. (**f**) A result page of a MAST analysis shows that three sequence motifs (red boxes) in a zebrafish sequence were found to be similar to the query motif (motif 4, inset). (**g**) A result page of a MATCH analysis in TRANSFAC shows that the tilapia *mpp5b* enhancer is predicted to contain binding sites for known transcription factors conservation of the TSS may suggest the TSS of medaka *mpp5b* to be 3 and 4 bp upstream of the FS519423 and FS516292, respectively. Despite this uncertainty, the 5′ end EST sequences provide a very close estimate of the TSS, thus revealing the region of the core promoter.

3. Identify conserved noncoding regions among orthologous sequences: Once the TSS has been estimated or pinpointed, the conserved upstream and downstream noncoding regions are revealed by high values in the PhastCons histograms of the evolutionary conservation and Multiz alignments of the five fish species of medaka, tetraodon, fugu, stickleback, and zebrafish on the UCSC Genome Browser. (Please note that coding regions are also highly conserved, so the conserved noncoding regions need to be distinguished from the coding regions, which normally align with the query amino acid sequences.) The conserved noncoding regions within 100 bp upstream and downstream of TSSs likely contain the core promoter; by contrast, more distal conserved noncoding regions, particularly in a single stretch of 80–250 bp, whether upstream or downstream of the TSS, likely harbor enhancers.

   For example, the PhastCons Conservation histograms and Multiz alignments revealed that, within 150 bp of the TSSs of teleost *mpp5b* genes (from −45 to +105), four short sequence motifs are conserved among medaka, stickleback, tetraodon, and fugu (please note that zebrafish sequence could not be aligned with other sequences at the TSS region, likely due to divergence in sequence and motif configuration) (Fig. 2c). Together, these four short motifs likely constitute the core promoter of the *mpp5b* genes. In addition, PhastCons Conservation and Multiz alignments also revealed conserved noncoding regions in the upstream intergenic and downstream intronic regions, which are candidates for the enhancer of *mpp5b* (Fig. 2b for two examples).

   Once conserved noncoding regions are identified, their sequences can be downloaded by **step 4**, their sequence motifs can be analyzed by **step 5**, and the possession of binding sites for known transcription factors can be assessed by **step 7** in Subheading 4.1.2. But if the corresponding conserved noncoding regions of the zebrafish gene are not revealed by PhastCons Conservation and Multiz alignments, as in the case of *mpp5b*, follow **step 6** in Subheading 4.1.2 to identify these regions.

4. Download the sequences of the conserved noncoding regions: First, select the regions in the interactive page to zoom into the conserved noncoding regions, and then click on the black bars in the information track of "Multiz alignment of 5 species" (Fig. 2b, at the bottom) to open sequence alignments as well as

the links for downloading the sequences in the FASTA format (Fig. 2d, showing a section of the *mpp5b* intron 1).

5. Identify conserved sequence motifs: To search the conserved noncoding regions for sequence motifs, we recommend using the MEME algorithm (a component of the MEME suite; http://meme-suite.org/) [39, 40]. Open the MEME website at http://meme-suite.org/tools/meme, input the sequences of the orthologous conserved noncoding regions in the FASTA format, and then, in the advanced options, choose the ranges of motif lengths arbitrarily between 6 and 17 to perform a search. We recommend searching repeatedly with different motif lengths because transcription factors bind to short motifs of different lengths, and MEME does not detect motifs outside the selected length ranges. Besides identifying conserved motifs, MEME also determines the consensus sequences of the identified motifs, which can be downloaded in the "Minimal MEME" format (Fig. 2e, clicking on the horizontal arrows to download) for later MAST and FIMO analyses as described below.

6. Identify corresponding noncoding regions in a zebrafish gene: To determine if a zebrafish DNA sequence contains regions that correspond to the conserved noncoding regions identified in other teleost fish by PhastCons Conservation and Multiz alignments, one can use the MAST and FIMO algorithms to search a candidate DNA sequence for short conserved sequence motifs identified by MEME. MAST and FIMO algorithms are designed to determine the occurrence and location of query motifs in DNA sequences (components of MEME suite, http://meme-suite.org/) [41, 42]. If a small region in the DNA sequence (e.g., 80–250 bp) is clustered with multiple short motifs identified previously by MEME, this region is likely the equivalent conserved noncoding region in zebrafish.

   For example, to determine if the first intron of the zebrafish *mpp5b* gene contains a region that corresponds to the candidate intronic enhancer identified in other fish species by PhastCons Conservation and Multiz alignments at **step 3** in Subheading 4.1.2 (Fig. 2b, right red arrowhead), open the MAST page at http://meme-suite.org/tools/mast and enter the sequence of the first intron of the zebrafish *mpp5b* gene (genomic sequences can be downloaded at the UCSC Genome Browser by clicking on the "View" tab and then the "DNA" link) and then individual motifs identified by MEME in the "Minimal MEME" format to perform the analysis. Figure 2f reveals the example results of searching a region within the first intron of zebrafish *mpp5b* for a single motif, noting that three occurrences of the motif were identified (red boxes). To increase the chance of identifying a conserved noncoding region, we recommend first searching with highly conserved motifs to narrow down the target regions and then searching with less-conserved motifs within the narrower regions. This motif-searching analysis can also be performed with the FIMO algorithm (http://meme-suite.org/tools/fimo). Ultimately, an intronic region in the middle of the first intron of the zebrafish *mpp5b* gene was identified to contain multiple conserved sequence motifs which were identified in other fish species by MEME, and this region was eventually confirmed to harbor the enhancer of zebrafish *mpp5b* [16].

7. Evaluate the presence of binding sites for known transcription factors in conserved noncoding regions: Authentic CREs contain binding sites for transcription factors. Thus, the presence of binding sites for known transcription factors in conserved noncoding regions would strongly suggest that they are authentic CREs. To evaluate whether conserved noncoding regions contain potential transcription factor binding sites, one can use TRANSFAC Professional (http://gene-regulation.com/), whose database contains experimentally verified information on eukaryotic transcription factors. (An institutional registration of TRANSFAC is required to access their most updated database.)

   For example, to search for potential transcription factor binding sites in the intronic enhancer candidate of the medaka *mpp5b* gene, open "MATCH tool" on the TRANSFAC webpage, then enter the conserved intronic sequence as an input, and click on "start search" to reveal a cluster of motifs that are potential binding sites for some transcription factors (Fig. 2g). (Please note that whether or not these transcription factors actually bind to the intronic sequence needs to be determined by experiments.)

Through the above steps, candidate core promoters and enhancers are predicted in Phase I. Please keep in mind that prediction is not 100% accurate. Therefore, the predicted candidate core promoters and enhancers need to be experimentally validated in Phase II of this protocol.

4.2 Phase II: Experimental Validation of Core Promoters and Enhancers

4.2.1 Rationale

In Phase II, candidate core promoters and enhancers need to be validated experimentally for three reasons:

1. The bioinformatic algorithms used for CRE prediction are not perfect, and making things worse, these imperfect algorithms are based on our incomplete understanding of transcriptional regulation.
2. Conserved noncoding sequences do not always carry transcriptional activities.

3. Functionally conserved CREs do not always display conservation in sequence homology because transcription factors can bind to degenerative sequences and because the number, location, and orientation of transcription factor binding sites can be flexible, thus evading homology detection [6, 43].

The transcriptional activities of candidate core promoters and enhancers can be assessed in transgenic zebrafish in vivo by three steps (Fig. 3a):

- First, generate transgenic constructs that use candidate CREs to control the expression of a fluorescent protein reporter gene, such as a green fluorescent protein (GFP) gene.
- Second, inject the constructs into zebrafish embryos and raise the injected embryos to desired developmental stages.
- Third, examine the expression patterns of the reporter gene in these transgenic fish to assess the transcriptional activities of the candidate CREs.

Phase II of the protocol utilizes a variety of techniques of molecular and cellular biology, such as PCR, restriction digestion, DNA ligation, plasmid isolation, plasmid construction, *E. coli* transformation, immunohistochemistry, and microscopy. For detailed explanations of these basic techniques, please refer to *Molecular Cloning: A Laboratory Manual* [44] and *Immunohistochemistry: Basics and Methods* [45]. Here, we explain only the outline of Phase II and some precautions pertinent to the purpose of this protocol.

4.2.2 Procedures

1. The general elements of a transgenic reporter construct: For a GFP reporter gene to be specifically expressed, the transgene construct must have several essential elements besides a core promoter and an enhancer. These elements should be arranged from the upstream end to the downstream end in the following order: 5′ untranslated region (5′ UTR), Kozak sequence, start codon, the open reading frame (ORF) of GFP, stop codon, and 3′ untranslated region (3′ UTR, containing a polyadenylation signal) (Fig. 3b). Many GFP expression vectors have these elements, so they can be modified to make a transgene reporter construct. For the basic properties of these elements, please refer to textbooks *Levin's Gene XII* and *Gene Control* [2, 46]. Besides maintaining the basic properties of these elements, the following specific precautions can also be used:

   5′ UTR: We recommend using the 5′ UTR of the gene of interest as the 5′ UTR of the transgenic reporter gene because 5′ UTRs likely contain core promoter motifs. Of course, when performing CRE-swapping experiments (see Phase III), 5′ UTRs of other genes can be used to create various

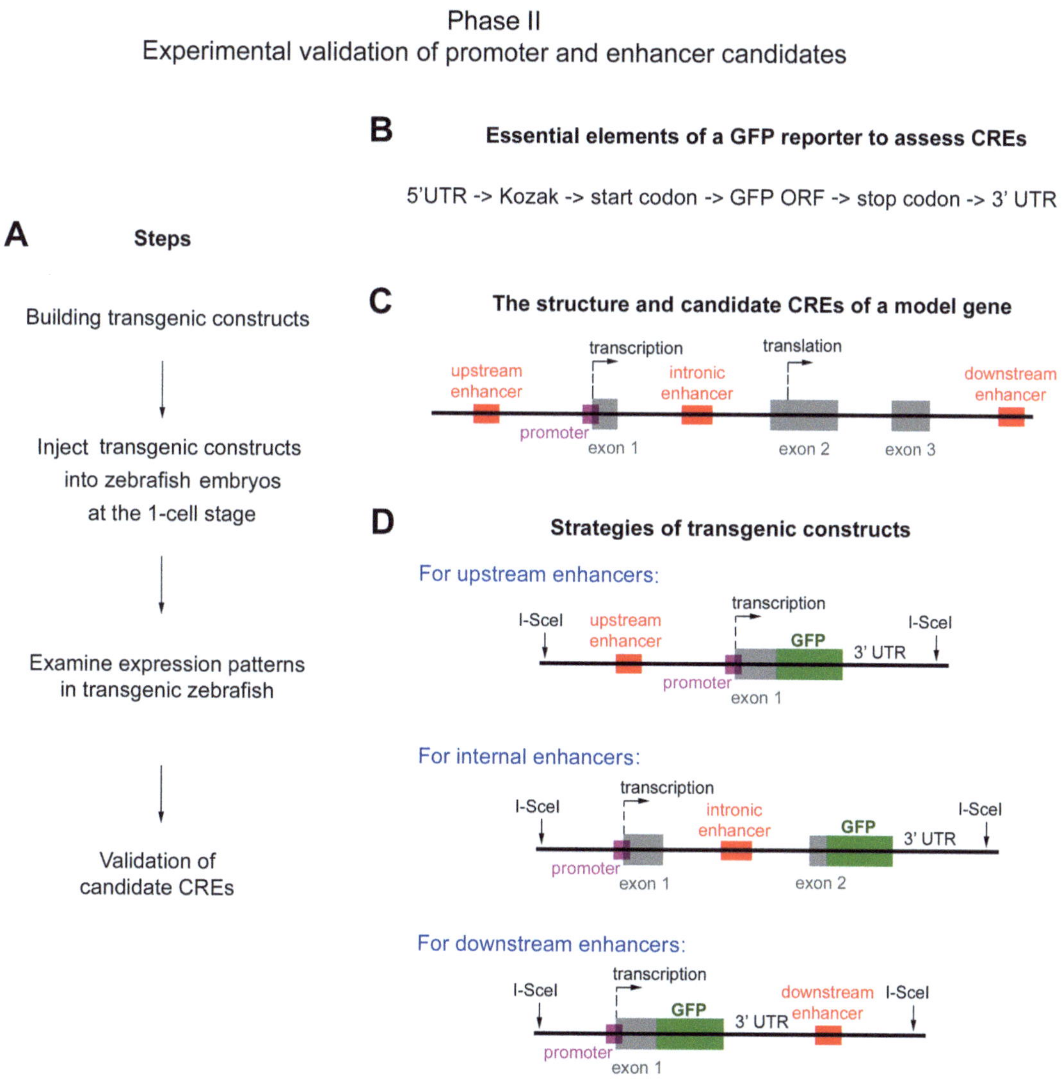

Fig. 3 The outline of Phase II of the protocol—experimental validation of core promoter and enhancer candidates. (**a**) The flowchart illustrates the steps to validate candidate core promoters and enhancers. (**b**) Some of the essential elements of a GFP reporter gene are presented from left to right in the upstream-to-downstream order in which they shall reside in the reporter gene. (**c**) The diagram illustrates the locations of the core promoter (a magenta box) and three possible locations of the enhancer of a model gene (red boxes, upstream, intronic, and downstream positions). Arrows indicate the transcription and translation start sites. Grey boxes stand for the exons of the model gene. (**d**) Diagrams illustrate three strategies for making transgene constructs to accommodate the three possible locations of enhancers

combinations of core promoters and enhancers to assess their transcriptional activities.

Kozak sequence: To ensure effective translation, a Kozak sequence (consensus sequence: GCCA/GCCATGG) needs to be included before the start codon. However, if the entire 5′ UTR of the gene of interest is to be included in a transgene construct, it is not necessary to include a synthetic Kozak sequence even if a Kozak consensus sequence is not apparent in the 5′ UTR because a functional substitute of the Kozak sequence should exist in the 5′ UTR.

3′ UTR: The SV40 3′ UTR can be used as the 3′ UTR for a reporter transgene because many GFP expression vectors use the SV40 3′ UTR, which contains a polyadenylation signal. Of course, the 3′ UTR of the gene of interest can also be used.

I-SceI sites: We recommend the I-SceI meganuclease-based transgenesis [17]. Thus, an I-SceI site needs to be placed at both the upstream and downstream boundaries of the transgene cassette (Fig. 3d).

2. Integrate candidate core promoters and enhancers into transgene constructs: Like the abovementioned essential elements of a gene, core promoters and enhancers also have their specific positions in a gene. The core promoter is positioned at the 5′ end of the gene. To ensure the entire core promoter is included in the construct, include some additional upstream sequence, 100–200 bp or thereabouts, along with conserved core promoter elements. (This precautious measure implies the possibility of including those enhancers that reside very close to TSSs in the core promoter candidates, thus making it more necessary to characterize CREs in the Phase III of the protocol to more accurately define the regions for the core promoter and enhancer.)

   Unlike core promoters, the enhancer of a gene can have three possible locations: upstream, intronic, and downstream (Fig. 3c). When integrating an enhancer candidate into a transgene construct, we recommend preserving its natural position and orientation because it is yet unclear what roles the location and orientation of enhancers play in their transcriptional activity, even though conventional wisdom, which is based on a limited number of case studies, suggests that enhancers can regulate transcriptional activities regardless of location and orientation [2]. Accordingly, to minimize the risk of overlooking an enhancer, three types of constructs are devised to test candidate enhancers at the three different locations (Fig. 3d). Of course, after an enhancer has been validated, it would be interesting to assess how the alteration of its location and orientation affects its transcriptional activities.

   When testing candidate enhancers, include some flanking non-conserved sequences as well so as to reduce the risk of missing less-conserved but functionally important motifs. Subsequently, the enhancer region can be trimmed by making deletion mutations with the Q5 Site-Directed Mutagenesis Kit (NEB, Cat #: E0554S). For example, when identifying the intronic enhancer of *mpp5b*, all of intron 1 was first assessed. Then un-conserved sequences were removed to narrow down the enhancer region [16]. When trimming an intronic enhancer, preserve the 5′ and 3′ splicing sites as well as the splicing branching site, which normally localizes 18–40 bases upstream of the 3′ splicing site [2].

3. Obtain the DNA fragments of candidate CREs: The sizes of candidate core promoters and enhancers are normally only a few hundred base pairs long. Thus, the DNA fragments of these candidate CREs can be amplified by PCR. High-fidelity DNA polymerase, such as the Platinum PCR SuperMix High-Fidelity DNA polymerase (Invitrogen, Cat #: 12532016), can be used for the PCR amplification. Template DNAs for PCR can be isolated from fish tissues with the Mini-Genomic DNA Buffer Set (QIAGEN, Cat #: 19060). Alternatively, if large pieces of DNA fragments are to be tested, they can also be isolated from a BAC or PAC DNA clone by restriction digestion or PCR amplification. BAC and PAC clones are available from the BACPAC Resources Center (https://bacpacresources.org/). BAC and PAC DNA can be purified with a QIAGEN Plasmid Mega Kit (Cat #: 12181).
4. Embryonic injection of transgene constructs: 50 pg of a transgenic construct can be coinjected with 0.1 unit of I-SceI meganuclease (NEB, R0694S) into zebrafish embryos at the one-cell stage. Detailed procedures of needle preparation and microinjection are illustrated by two video articles [18, 19]. After injection, raise the embryos in E3 medium (5 mM NaCl, 0.17 mM KCl, 0.33 mM $CaCl_2$, 0.33 mM $MgSO_4$, 0.00001% (w/v) methylene blue) until 5 days postfertilization (dpf). To examine GFP expression in the eyes, melanin pigmentation needs to be blocked with 0.0003% phenylthiourea (PTU) in the E3 medium. When necessary, transfer the fish larva to an aquatic system to raise to desired developmental stages, such as 9 dpf or adulthood. The GFP expression patterns of the fish can be analyzed by confocal immunohistochemistry as follows.
5. Evaluate GFP expression patterns by immunohistochemistry: Zebrafish retina has one type of rod photoreceptor and four types of cone photoreceptors (red, green, blue, and UV cones). The resulting GFP expression patterns in the photoreceptors of either the adult retina or the larval retina can be determined by confocal immunohistochemistry per the morphological characteristics and immunoreactivities as follows:

   In the adult zebrafish retina, five criteria can be used to distinguish the types of photoreceptors that express GFP:

   (a) The regular planar patterning of zebrafish photoreceptors. In zebrafish, green, red, and blue cones coalesce into

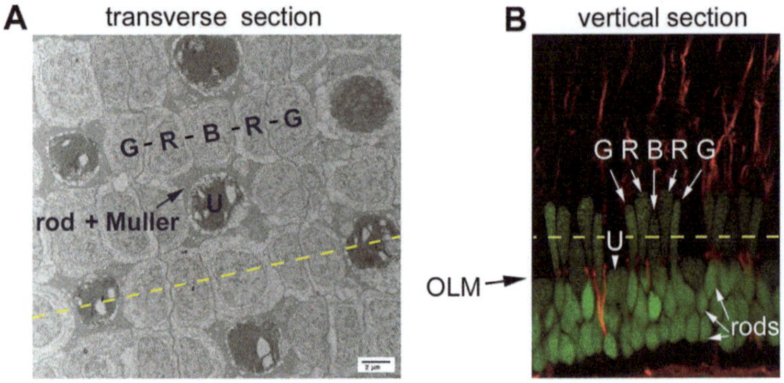

Fig. 4 Morphological characteristics of five types of zebrafish photoreceptors. (**a**) A TEM image of a transverse section of the photoreceptor cell layer of the adult zebrafish retina shows the mirror-symmetric pentameric alignment of red, green, and blue (RGB) cones, indicated by G-R-B-R-G. U stands for UV cones, which display darkly stained outer segments. An arrow indicates the clusters of the inner segments of rod photoreceptors and the apical processes of Müller glial cells around UV cones. The yellow dash line indicates the direction and location of vertical sectioning that would produce a retinal section as shown in panel **b**. (**b**) A confocal immunohistochemical image of a vertical section of the adult zebrafish retina shows the morphologies and patterning of photoreceptor nuclei (by DAPI staining, green). Note that the elongated RGB cone nuclei locate apical to the outer limiting membrane (OLM), the round nuclei of UV cones (U) localize immediately basal to the OLM, and the strongly stained rod nuclei locate away from OLM in the basal half of the outer nuclear layer. Red signals show the counterstaining of acetylated tubulin, some of which illustrate the OLM region. The yellow dash line indicates the direction and position of transverse sectioning that would produce a retinal section as shown in panel (**a**)

pentamers in the order of G-R-B-R-G within each row of photoreceptors, with a UV cone separating two cone pentamers [47–49]; between neighboring rows, the positions of pentamers shift by three cells, generating alternating columns of blue-UV cones and columns of red-green cones. Unlike cones, the inner segments of rod photoreceptors cluster around UV cones, and the cross sections of their inner segments are very thin at the level of RGB cone nuclei (Fig. 4a; [15, 50]; Table 1).

(b) The DAPI staining of the nuclei of RGB cones and UV cone is lighter than that of rod nuclei (Fig. 4b; Table 1).

(c) The nuclei of different types of photoreceptor localize to distinct regions relative to the outer limiting membrane (OLM), with rod nuclei basal and distal to the OLM, UV cone nuclei basal to and juxtaposing the OLM, and elongated RGB cone nuclei apical to and juxtaposing the OLM (Fig. 4b; Table 1).

Table 1
The table summarizes the morphological and immunoreactive characteristics of adult zebrafish photoreceptors

Characteristics	Green	Red	Blue	UV	Rod
As pentamers	+	+	+	–	–
Apical to OLM	+	+	+	–	–
Dense DAPI	–	–	–	–	+
Zpr1	+	+	–	–	–
Green opsin	+	–	–	–	–
Red opsin	–	+	–	–	–
Blue opsin	–	–	+	–	–
UV opsin	–	–	–	+	–
Rod opsin	–	–	–	–	+
Zpr3	–	–	–	–	+

Plus signs stand for possession of the characteristics and minus signs for nonpossession

(d) Immunoreactivity to photoreceptor type-specific antibodies (Table 1, the opsin antibodies, [51]; Zpr1 antibody for double cones, red and green cones, ZFIN; Zpr3 antibody for rods, ZFIN).

(e) Comparison with existing transgenic zebrafish lines, in which various types of photoreceptors are highlighted by transgenic proteins (Table 2 and references therein) [16, 50, 52–61].

In the larval zebrafish retina, cones do not align in regular mosaics as in the adult retina [62]. Thus, the planar arrangement of cones in the larval retina cannot be used as a criterion to distinguish photoreceptor types. However, the intensity of nuclear staining, immunoreactivities, and comparison with existing transgenic zebrafish lines can still be utilized as criteria to distinguish photoreceptor types such as for the adult retina (Tables 2 and 3). In addition, the following morphological aspects of larval photoreceptors differ among photoreceptor types and can also be used as photoreceptor-type-identifying criteria: the positions of the nuclei, the position of the ellipsoid, and the cross-sectional sizes of inner segments at the OLM (Table 3) [16].

Although the above procedures are tailored for validating the transcriptional activity and specificity of the candidate core promoter and enhancer of a photoreceptor type-specific gene, by modifying the methods and criteria to assess GFP expression patterns, the procedure can also be used to validate the transcriptional

Table 2
The table summarizes the expression patterns of fluorescent reporter proteins in the photoreceptors of some published transgenic zebrafish lines

Transgenic lines	Red	Green	Blue	UV	Rod	References
Tg(LWS1/GFP-LWS2/RFP-PAC(E))	+	−	−	−	−	Tsujimura et al. (2010) [52]
Tg(LWS)	+	−	−	−	−	Crespo et al. (2018) [53]
Tg(LCR^{RH2}-RH2-2:GFP)$^{pt115\text{-}k}$	−	+	−	−	−	Fang et al. (2013) [50]
Tg(RH2-1/GFP-PAC)	−	+	−	−	−	Tsujimura et al. (2007) [54]
Tg(RH2-1:HA-mCherry)^{pt120}	−	+	−	−	−	Fang et al. (2017) [16]
Tg(zfSWS2-1.1A:EGFP)	−	−	+	−	−	Takechi et al. (2008) [55]
Tg(SWS1:GFP)	−	−	−	+	−	Takechi et al. (2003) [56]
Tg(SWS1:GFP)	−	−	−	+	−	Luo et al. (2004) [57]
Tg(LCR^{RH2}-RH2-1:GFP)^{pt112}	−	+	+	−	−	Fang et al. (2013) [50]
Tg(ponli^{6,102} :HA-mCherry)^{pt118b}	+	+	+	−	−	Fang et al. (2017) [16]
TG(3.2T CP-EGFP)	+	+	+	+	−	Kennedy et al. (2007) [58]
Tg(-3.7rho:EGFP)^{kj2}	−	−	−	−	+	Hamaoka et al. (2002) [59]
Tg(XOPS-EGFP)	−	−	−	−	+	Fadool (2003) [60]
Tg(pZOP-EGFP)	−	−	−	−	+	Kennedy et al. (2001) [61]

Table 3
The table summarizes the morphological and immunoreactive characteristics of larval zebrafish photoreceptors at 9 dpf

Characteristics	Green	Red	Blue	UV	Rod
Nuclei crossing OLM	+	+	+	−	−
Dense DAPI	−	−	−	−	+
Zpr1	+	+	−	−	−
Green opsin	+	−	−	−	−
Red opsin	−	+	−	−	−
Blue opsin	−	−	+	−	−
UV opsin	−	−	−	+	−
Rod opsin	−	−	−	−	+
Zpr3	−	−	−	−	+
Cross-section size at the OLM	Large	Large	Large	Large	Small
Ellipsoid distance to the OLM	Far	Far	Far	Close	Far

Plus signs stand for possession of the characteristics and minus signs for nonpossession activity of the CREs of other tissue-specific or cell type-specific genes.

4.3 Characterization of CREs

4.3.1 Rationale

Once the core promoter and enhancer regions of a gene are validated in Phase II, these CRE regions can be further characterized in Phase III. For example, by swapping CREs, deleting CREs, or mutating CRE motifs, one can further trim and narrow down the CRE regions, evaluate the contributions of individual CREs or CRE motifs to the transcriptional specificity of the gene, determine the functional conservation among orthologous CREs, or infer the grammatical code of the CREs. These analyses are explained as follows.

4.3.2 Procedures

1. CRE-swapping analysis: One way to dissect the functions of core promoters and enhancers is to analyze the expression patterns of reporter genes that are driven by the combination of the core promoter of one gene and the enhancer of another gene. The expression patterns driven by such chimeric combinations of CREs of two genes can then be compared with those by the CREs of a single gene to evaluate the roles of individual CREs in transcription.

   For example, when the combination of the core promoter of the broadly expressing *mpp5a* gene (also known as *nagie oko*) [63] and the enhancer of RGB cone-specific *mpp5b* gene drove RGB cone-specific transcription, we could infer that the *mpp5b* enhancer plays a critical role in RGB cone-specific transcription [16]. Similarly, the RGB cone-specific transcription driven by the combination of zebrafish *mpp5b* core promoter and tilapia *mpp5b* enhancer suggests that the enhancers of *mpp5b* orthologs are conserved among teleost species [16].

2. Deletion and mutation analyses: More drastic than the CRE-swapping analysis, deletion analysis can be performed to remove or trim the core promoter and the enhancer of a transgenic GFP reporter construct. The resulting effects on transcriptional activity will shed light on the functions of the deleted regions. Moreover, the individual CRE motifs can be mutated to evaluate the roles of each motif in transcription. When mutating individual motifs, it is recommended that a motif be replaced with an unrelated sequence of the same length (viz., making substitution mutations rather than deletion mutations) because if the length of the CRE is altered by deletion, the spatial orientations of the neighboring binding sites for transcription factors and their intervals may be altered, potentially disrupting transcription factor interactions required for transcription, even if the motif sequence per se does not play a role in transcription.

   When making a substitution mutation, a restriction enzyme site can be embedded in the substituting sequence. The inclusion of such an enzyme site allows for easy confirmation of the mutation by restriction digestion. Both deletion and substitution mutations can be achieved with the Q5 Site-Directed Mutagenesis Kit (NEB, Cat #: E0554S).

Acknowledgments

This work is supported by the National Institutes of Health (P30EY008098; EY025638; R21EY023665) as well as by the grants to the Department of Ophthalmology of the University of Pittsburgh from the Eye and Ear Foundation of Pittsburgh and Research to Prevent Blindness. The authors declare no competing financial interests. We thank Ms. Lynne Sunderman for proofreading the manuscript.

References

1. Hughes TR (2011) A handbook of transcription factors. Springer, Dordrecht Heidelberg
2. Latchman D (2015) Gene control. Garland Science, New York
3. Latchman D (2008) Eukaryotic transcription factors, 5th edn. Academic Press, London
4. Juven-Gershon T, Kadonaga JT (2010) Regulation of gene expression via the core promoter and the basal transcriptional machinery. Dev Biol 339(2):225–229. https://doi.org/10.1016/j.ydbio.2009.08.009
5. Yanez-Cuna JO, Kvon EZ, Stark A (2013) Deciphering the transcriptional cis-regulatory code. Trends Genet 29(1):11–22. https://doi.org/10.1016/j.tig.2012.09.007
6. Nelson AC, Wardle FC (2013) Conserved non-coding elements and cis regulation: actions speak louder than words. Development 140(7):1385–1395. https://doi.org/10.1242/dev.084459
7. Long HK, Prescott SL, Wysocka J (2016) Ever-changing landscapes: transcriptional enhancers in development and evolution. Cell 167(5):1170–1187. https://doi.org/10.1016/j.cell.2016.09.018
8. Lenhard B, Sandelin A, Carninci P (2012) Metazoan promoters: emerging characteristics and insights into transcriptional regulation. Nat Rev Genet 13(4):233–245. https://doi.org/10.1038/nrg3163
9. Haberle V, Stark A (2018) Eukaryotic core promoters and the functional basis of transcription initiation. Nat Rev Mol Cell Biol 19 (10):621–637. https://doi.org/10.1038/s41580-018-0028-8
10. Kulaeva OI, Nizovtseva EV, Polikanov YS, Ulianov SV, Studitsky VM (2012) Distant activation of transcription: mechanisms of enhancer action. Mol Cell Biol 32 (24):4892–4897. https://doi.org/10.1128/MCB.01127-12
11. Lodish H, Berk A, Zipursky SL, Matsudaira P, Baltimore D, Darnell J (2000) Regulatory sequences in eukaryotic protein-coding genes. Molecular cell biology, 4th edn. W. H. Freeman and Company, New York
12. Bulger M, Groudine M (1999) Looping versus linking: toward a model for long-distance gene activation. Genes Dev 13(19):2465–2477. https://doi.org/10.1101/gad.13.19.2465
13. Chong JA, Tapia-Ramirez J, Kim S, Toledo-Aral JJ, Zheng Y, Boutros MC, Altshuller YM, Frohman MA, Kraner SD, Mandel G (1995) REST: a mammalian silencer protein that restricts sodium channel gene expression to neurons. Cell 80(6):949–957. https://doi.org/10.1016/0092-8674(95)90298-8
14. Herold M, Bartkuhn M, Renkawitz R (2012) CTCF: insights into insulator function during development. Development 139 (6):1045–1057. https://doi.org/10.1242/dev.065268
15. Zou J, Yang X, Wei X (2010) Restricted localization of ponli, a novel zebrafish MAGUK-family protein, to the inner segment interface areas between green, red, and blue cones. Invest Ophthalmol Vis Sci 51(3):1738–1746. https://doi.org/10.1167/iovs.09-4520
16. Fang W, Guo C, Wei X (2017) Rainbow enhancers regulate restrictive transcription in teleost green, red, and blue cones. J Neurosci 37 (11):2834–2848. https://doi.org/10.1523/JNEUROSCI.3421-16.2017
17. Thermes V, Grabher C, Ristoratore F, Bourrat F, Choulika A, Wittbrodt J, Joly JS

(2002) I-SceI meganuclease mediates highly efficient transgenesis in fish. Mech Dev 118 (1–2):91–98

18. Yuan S, Sun Z (2009) Microinjection of mRNA and morpholino antisense oligonucleotides in zebrafish embryos. J Vis Exp (27). https://doi.org/10.3791/1113
19. Rosen JN, Sweeney MF, Mably JD (2009) Microinjection of zebrafish embryos to analyze gene function. J Vis Exp (25). https://doi.org/10.3791/1115
20. Hardison RC, Taylor J (2012) Genomic approaches towards finding cis-regulatory modules in animals. Nat Rev Genet 13 (7):469–483. https://doi.org/10.1038/nrg3242
21. Hardison RC (2000) Conserved noncoding sequences are reliable guides to regulatory elements. Trends Genet 16(9):369–372
22. Pennacchio LA, Rubin EM (2001) Genomic strategies to identify mammalian regulatory sequences. Nat Rev Genet 2(2):100–109. https://doi.org/10.1038/35052548
23. Vavouri T, Lehner B (2009) Conserved noncoding elements and the evolution of animal body plans. BioEssays 31(7):727–735. https://doi.org/10.1002/bies.200900014
24. Elgar G (2009) Pan-vertebrate conserved non-coding sequences associated with developmental regulation. Brief Funct Genomic Proteomic 8(4):256–265. https://doi.org/10.1093/bfgp/elp033
25. Goodrich JA, Tjian R (2010) Unexpected roles for core promoter recognition factors in cell-type-specific transcription and gene regulation. Nat Rev Genet 11(8):549–558. https://doi.org/10.1038/nrg2847
26. Panne D, Maniatis T, Harrison SC (2007) An atomic model of the interferon-beta enhanceosome. Cell 129(6):1111–1123. https://doi.org/10.1016/j.cell.2007.05.019
27. Nikolov DB, Hu SH, Lin J, Gasch A, Hoffmann A, Horikoshi M, Chua NH, Roeder RG, Burley SK (1992) Crystal structure of TFIID TATA-box binding protein. Nature 360(6399):40–46. https://doi.org/10.1038/360040a0
28. Chalkley GE, Verrijzer CP (1999) DNA binding site selection by RNA polymerase II TAFs: a TAF(II)250-TAF(II)150 complex recognizes the initiator. EMBO J 18(17):4835–4845. https://doi.org/10.1093/emboj/18.17.4835
29. Beer MA, Tavazoie S (2004) Predicting gene expression from sequence. Cell 117 (2):185–198. https://doi.org/10.1016/s0092-8674(04)00304-6
30. Yuan Y, Guo L, Shen L, Liu JS (2007) Predicting gene expression from sequence: a reexamination. PLoS Comput Biol 3(11):e243. https://doi.org/10.1371/journal.pcbi.0030243
31. Middendorf M, Kundaje A, Wiggins C, Freund Y, Leslie C (2004) Predicting genetic regulatory response using classification. Bioinformatics 20(Suppl 1):i232–i240. https://doi.org/10.1093/bioinformatics/bth923
32. Warner JB, Philippakis AA, Jaeger SA, He FS, Lin J, Bulyk ML (2008) Systematic identification of mammalian regulatory motifs' target genes and functions. Nat Methods 5 (4):347–353. https://doi.org/10.1038/nmeth.1188
33. Rouault H, Santolini M, Schweisguth F, Hakim V (2014) Imogene: identification of motifs and cis-regulatory modules underlying gene co-regulation. Nucleic Acids Res 42 (10):6128–6145. https://doi.org/10.1093/nar/gku209
34. Tagle DA, Koop BF, Goodman M, Slightom JL, Hess DL, Jones RT (1988) Embryonic epsilon and gamma globin genes of a prosimian primate (Galago crassicaudatus). Nucleotide and amino acid sequences, developmental regulation and phylogenetic footprints. J Mol Biol 203(2):439–455. https://doi.org/10.1016/0022-2836(88)90011-3
35. Das MK, Dai HK (2007) A survey of DNA motif finding algorithms. BMC Bioinformatics 8(Suppl 7):S21. https://doi.org/10.1186/1471-2105-8-S7-S21
36. Kent WJ (2002) BLAT—the BLAST-like alignment tool. Genome Res 12(4):656–664. https://doi.org/10.1101/gr.229202
37. Kent WJ, Sugnet CW, Furey TS, Roskin KM, Pringle TH, Zahler AM, Haussler D (2002) The human genome browser at UCSC. Genome Res 12(6):996–1006. https://doi.org/10.1101/gr.229102
38. Carninci P, Sandelin A, Lenhard B, Katayama S, Shimokawa K, Ponjavic J, Semple CA, Taylor MS, Engstrom PG, Frith MC, Forrest AR, Alkema WB, Tan SL, Plessy C, Kodzius R, Ravasi T, Kasukawa T, Fukuda S, Kanamori-Katayama M, Kitazume Y, Kawaji H, Kai C, Nakamura M, Konno H, Nakano K, Mottagui-Tabar S, Arner P, Chesi A, Gustincich S, Persichetti F, Suzuki H, Grimmond SM, Wells CA, Orlando V, Wahlestedt C, Liu ET, Harbers M, Kawai J, Bajic VB, Hume DA, Hayashizaki Y (2006) Genome-wide analysis of mammalian promoter architecture and evolution. Nat Genet 38(6):626–635. https://doi.org/10.1038/ng1789

39. Bailey TL, Boden M, Buske FA, Frith M, Grant CE, Clementi L, Ren J, Li WW, Noble WS (2009) MEME SUITE: tools for motif discovery and searching. Nucleic Acids Res 37(Web Server):W202–W208. https://doi.org/10.1093/nar/gkp335
40. Bailey TL, Elkan C (1994) Fitting a mixture model by expectation maximization to discover motifs in biopolymers. Proc Int Conf Intell Syst Mol Biol 2:28–36
41. Bailey TL, Gribskov M (1998) Combining evidence using p-values: application to sequence homology searches. Bioinformatics 14(1):48–54. https://doi.org/10.1093/bioinformatics/14.1.48
42. Grant CE, Bailey TL, Noble WS (2011) FIMO: scanning for occurrences of a given motif. Bioinformatics 27(7):1017–1018. https://doi.org/10.1093/bioinformatics/btr064
43. Wittkopp PJ, Kalay G (2011) Cis-regulatory elements: molecular mechanisms and evolutionary processes underlying divergence. Nat Rev Genet 13(1):59–69. https://doi.org/10.1038/nrg3095
44. Green M, Sambrook J (2012) Molecular cloning: a laboratory manual, 4th edn. CSHL Press, New York
45. Buchwalow I, Böcker W (2010) Immunohistochemistry: basics and methods. Springer, Berlin, Heidelberg
46. Krebs J, Goldstein E, Kilpatrick S (2017) Levin's Gene XII. Jones & Bartlett Learning, Burlington
47. Raymond PA, Barthel LK, Curran GA (1995) Developmental patterning of rod and cone photoreceptors in embryonic zebrafish. J Comp Neurol 359(4):537–550. https://doi.org/10.1002/cne.903590403
48. Robinson J, Schmitt EA, Harosi FI, Reece RJ, Dowling JE (1993) Zebrafish ultraviolet visual pigment: absorption spectrum, sequence, and localization. Proc Natl Acad Sci U S A 90(13):6009–6012. https://doi.org/10.1073/pnas.90.13.6009
49. Zou J, Wang X, Wei X (2012) Crb apical polarity proteins maintain zebrafish retinal cone mosaics via intercellular binding of their extracellular domains. Dev Cell 22(6):1261–1274. https://doi.org/10.1016/j.devcel.2012.03.007
50. Fang W, Bonaffini S, Zou J, Wang X, Zhang C, Tsujimura T, Kawamura S, Wei X (2013) Characterization of transgenic zebrafish lines that express GFP in the retina, pineal gland, olfactory bulb, hatching gland, and optic tectum. Gene Expr Patterns 13(5–6):150–159. https://doi.org/10.1016/j.gep.2013.02.006
51. Vihtelic TS, Doro CJ, Hyde DR (1999) Cloning and characterization of six zebrafish photoreceptor opsin cDNAs and immunolocalization of their corresponding proteins. Vis Neurosci 16(3):571–585
52. Tsujimura T, Hosoya T, Kawamura S (2010) A single enhancer regulating the differential expression of duplicated red-sensitive opsin genes in zebrafish. PLoS Genet 6(12):e1001245. https://doi.org/10.1371/journal.pgen.1001245
53. Crespo C, Soroldoni D, Knust E (2018) A novel transgenic zebrafish line for red opsin expression in outer segments of photoreceptor cells. Dev Dyn 247(7):951–959. https://doi.org/10.1002/dvdy.24631
54. Tsujimura T, Chinen A, Kawamura S (2007) Identification of a locus control region for quadruplicated green-sensitive opsin genes in zebrafish. Proc Natl Acad Sci U S A 104(31):12813–12818. https://doi.org/10.1073/pnas.0704061104
55. Takechi M, Seno S, Kawamura S (2008) Identification of cis-acting elements repressing blue opsin expression in zebrafish UV cones and pineal cells. J Biol Chem 283(46):31625–31632. https://doi.org/10.1074/jbc.M806226200
56. Takechi M, Hamaoka T, Kawamura S (2003) Fluorescence visualization of ultraviolet-sensitive cone photoreceptor development in living zebrafish. FEBS Lett 553(1–2):90–94. https://doi.org/10.1016/s0014-5793(03)00977-3
57. Luo W, Williams J, Smallwood PM, Touchman JW, Roman LM, Nathans J (2004) Proximal and distal sequences control UV cone pigment gene expression in transgenic zebrafish. J Biol Chem 279(18):19286–19293. https://doi.org/10.1074/jbc.M400161200
58. Kennedy BN, Alvarez Y, Brockerhoff SE, Stearns GW, Sapetto-Rebow B, Taylor MR, Hurley JB (2007) Identification of a zebrafish cone photoreceptor-specific promoter and genetic rescue of achromatopsia in the nof mutant. Invest Ophthalmol Vis Sci 48(2):522–529. https://doi.org/10.1167/iovs.06-0975
59. Hamaoka T, Takechi M, Chinen A, Nishiwaki Y, Kawamura S (2002) Visualization of rod photoreceptor development using GFP-transgenic zebrafish. Genesis 34(3):215–220. https://doi.org/10.1002/gene.10155

60. Fadool JM (2003) Development of a rod photoreceptor mosaic revealed in transgenic zebrafish. Dev Biol 258(2):277–290. https://doi.org/10.1016/s0012-1606(03)00125-8
61. Kennedy BN, Vihtelic TS, Checkley L, Vaughan KT, Hyde DR (2001) Isolation of a zebrafish rod opsin promoter to generate a transgenic zebrafish line expressing enhanced green fluorescent protein in rod photoreceptors. J Biol Chem 276(17):14037–14043. https://doi.org/10.1074/jbc.M010490200
62. Allison WT, Barthel LK, Skebo KM, Takechi M, Kawamura S, Raymond PA (2010) Ontogeny of cone photoreceptor mosaics in zebrafish. J Comp Neurol 518 (20):4182–4195. https://doi.org/10.1002/cne.22447
63. Wei X, Malicki J (2002) Nagie oko, encoding a MAGUK-family protein, is essential for cellular patterning of the retina. Nat Genet 31 (2):150–157. https://doi.org/10.1038/ng883

Chapter 11

Ultrasensitive RNAscope In Situ Hybridization System on Embryonic and Adult Mouse Retinas

Takae Kiyama and Chai-An Mao

Abstract

In situ hybridization (ISH) techniques provide important information regarding gene expression in cells and tissues. Especially, ISH details complex spatial RNA expression in highly heterogeneous tissues, such as developing and mature central nervous systems, where rare genes involved in many fundamental developmental or biological events are expressed. Although several techniques have been developed to detect low levels of RNA expression, there are still problematic issues caused by a low signal-to-noise ratio after signal amplification. RNAscope is a recently developed ISH technique with high sensitivity and low background. RNAscope utilizes a unique probe system (double Z probe) to amplify signal from rare RNAs. Additionally, the double Z probe enables a significant reduction in nonspecific signal amplification. Here we report detailed procedures of the brown-color RNAscope ISH on embryonic and adult mouse retinas.

Key words In situ hybridization, RNAscope, Mouse retina, FFPE section, Brown-color reaction, RNA, Gene expression

1 Introduction

The central nervous system is a highly organized structure composed of heterogeneous types of cells. These different types or subtypes of cells express various combinations of genes, enabling complex developmental and neural functions. To gain a comprehensive understanding of the function of these genes, it is important to understand the precise spatial and temporal gene expression.

In situ hybridization, a technique to detect RNA expression using antisense RNA probes, has been used and refined for decades to detect spatiotemporal gene expression patterns [1]. The first ISH technique used isotopically labeled probes. Isotopic ISH was sensitive but required long antisense RNA probes, which caused high background noise due to nonspecific hybridization and long exposure time. Subsequently, biotin-labeled oligonucleotide probes and fluorescent-labeled oligonucleotide probes were invented to detect RNA using chromogenic reactions or

Chai-An Mao (ed.), *Retinal Development: Methods and Protocols*, Methods in Molecular Biology, vol. 2092,
https://doi.org/10.1007/978-1-0716-0175-4_11,

fluorescence, avoiding the difficulties of using radioisotopes [2, 3]. Although these probe systems improved ISH techniques and shortened the procedure time, they lacked sensitivity for low gene expression. Additionally, the signal-to-noise ratio was limited because of nonspecific hybridization. To improve the sensitivity, in situ PCR, which amplifies target mRNAs before hybridization, and signal amplification methods after hybridization were developed [4–6]. However, despite these technical improvements, there were still problematic issues caused by amplifying signal and background noise to detect low levels of RNA. Currently, RNA probes are most widely used for ISH as they are able to provide stable hybridization with target RNA. However, because RNA probes are very sensitive to RNase, probes can be easily destroyed. To prevent probe degradation, a scrupulously RNase-free environment is required. In addition, due to its long size (100s bases), RNA probes require more time for penetration. Also, the long probes may cause nonspecific hybridization.

In this chapter, we describe the application of RNAscope, an extremely sensitive ISH technology, in determining the spatiotemporal gene expression patterns in mouse retinas. The RNAscope ISH technique uses oligonucleotide probes that employ a branched DNA (bDNA) signal amplification assay to detect target-specific signals with low background [6]. A standard RNAscope ISH probe is composed of ~20 target-gene-specific double Z probes. Each pair of the target-specific double Z probes first hybridizes to the target RNA molecule in tandem to ensure target-specific signal amplification. Subsequently, the amplifiers and detection probes (containing either a fluorescent molecule or a chromogenic reagent) will bind to the double Z probes, amplifying the signal and resulting in ultrasensitive and high signal-to-noise RNA detection [7] (*see* **Note 1**).

RNAscope enables ISH to be performed easier and faster with high reproducibility for daily use. Oligonucleotide Z probes are stable and do not require RNase-free sterilization. Additionally, the small size of the RNAscope probes allows for tissue penetration in a shorter amount to time than traditional RNA probes, resulting in a much shorter experimental time. Once the sample sections are ready, a full ISH experiment can be done within 2 days. Each oligonucleotide probe is designed to have similar parameters, such as GC contents, Tm value, and probe concentration, to allow for highly reproducible results in different tissue samples and with various detection methods.

RNAscope ISH methods have been cited in more than 1000 publications in a broad range of research fields, including developmental biology, neuroscience, and cancer, and using a variety of species and tissues. The double Z designed probe can be utilized for both fluorescence and color detection ISH using formalin-fixed paraffin-embedded (FFPE) or cryo-sectioned retinal samples [8–10]. Furthermore, RNAscope ISH can be used in conjunction with immunofluorescent staining to visualize both mRNAs and protein localization in mouse retina [11]. It has also been used on whole-mount zebrafish and chicken embryos and can be used on flat-mounted retinas as well [12, 13] (*see* **Note 2**). Here we outline a protocol for RNAscope ISH using FFPE-sectioned adult and embryonic mouse retinas with brown-color reaction.

2 Materials

2.1 RNAscope Reagents

1. RNAscope 2.5 HD Detection Reagent-BROWN: Store at 4 °C.
2. Target probes (color channel 1): Store at 4 °C.
3. Proteinase III: Store at 4 °C.
4. Target Retrieval Reagents: Prepare 1× target retrieval reagent by adding 10 mL of 10× target retrieval reagent to 90 mL of DEPC-treated water (DEPC-DW).
5. Wash buffer: Prepare 200 mL of 1× wash buffer by adding 4 mL of 50× wash buffer to 180 mL of distilled water. Mix well.

2.2 Equipment

1. HybEZ Hybridization system: HybEZ Oven and HybEZ Humidity Control Tray with lid are used for hybridization and color reaction with the RNAscope system.
2. Vacuum oven to keep temperature at 65 °C for embedding paraffin sample.
3. Microtome (ThermoFisher Scientific) to section FFPE sample.
4. Water bath to keep temperature at 40 °C for FFPE sectioning.
5. Heating plate (VWR) for pretreatment target retrieval.

2.3 Other Materials and Reagents

2.3.1 Fixation, Dehydration, Embedding, and Sectioning

1. 10% neutral buffered formalin.
2. Phosphate buffered saline (PBS).
3. 30%, 50%, 70%, and 100% ethanol (EtOH).
4. Xylene.
5. Paraffin wax.
6. SuperFrost Plus Slides (Fisher Scientific).
7. DEPC-DW.

2.3.2 Pretreatment

1. Hydrogen peroxide (Mallinckrodt Chemicals): Prepare 3% hydrogen peroxide solution by adding 5 mL of 30% of hydrogen peroxide to 45 mL of DEPC-DW. Mix well.
2. ImmEdge hydrophobic pen (Vector Laboratories).

2.3.3 Counterstain and Mount

1. Mayer's hematoxylin (Sigma-Aldrich): Prepare 50% hematoxylin solution by diluting with distilled water.
2. Ammonium hydroxide solution (Sigma-Aldrich): Prepare 0.02% (w/v) ammonia and water by adding 143 μL of 1 N ammonium hydroxide to 25 mL of distilled water. Mix well.
3. Cytoseal 60 (Richard-Allan Scientific).
4. Cover glass 24 × 50 mm (Electron Microscopy Science).

3 Methods

3.1 Fixation and Dehydration

1. Extract mouse embryos from the uterus and place in ice-cold PBS. To avoid cross contamination, embryos should be placed separately. Place three to four embryos in a 10 cm petri dish.
2. Place one embryo in a clean petri dish under a dissection microscope with clean forceps. Snip a small piece of tissue (e.g., half of the tail at E15.5) and place in a clean microcentrifuge tube for genotyping. In between cutting each embryo, wash forceps thoroughly with running water to avoid cross contamination of samples. Cut the embryo head off with a razor blade (*see* **Note 3**). If using adult eyes, dissect the eyeballs from a euthanized mouse. Fix embryonic heads or eyeballs in 10% neutral buffered formalin in 5 mL test tube (round bottom) at 4 °C overnight.
3. During fixation, proceed with genotyping.
4. Wash samples with PBS three times for 20 min each at room temperature with gentle rocking.
5. Dehydrate samples sequentially through 30%, 50%, 70%, and 100% (twice) EtOH at room temperature with gentle rocking (*see* **Note 4**). Dehydrated samples can be stored in 100% EtOH at −80 °C for at least 6 months.

3.2 Embedding

1. Transfer samples under xylene in a glass scintillation vial. Keep at room temperature with gentle rocking for 3 h. Change to fresh xylene and keep at room temperature with gentle rocking for another 5 h.
2. Remove xylene and add melted paraffin. Loosen the lid of the scintillation vial and place the vial in a vacuum oven at 65 °C overnight. Discard the paraffin the next morning, replace with new paraffin, and leave vials in the vacuum oven at 65 °C for 3 h. Repeat cycle one additional time.
3. Embed sample in mold. Position the sample and mark the direction. Embedded sample can be stored at 4 °C indefinitely.

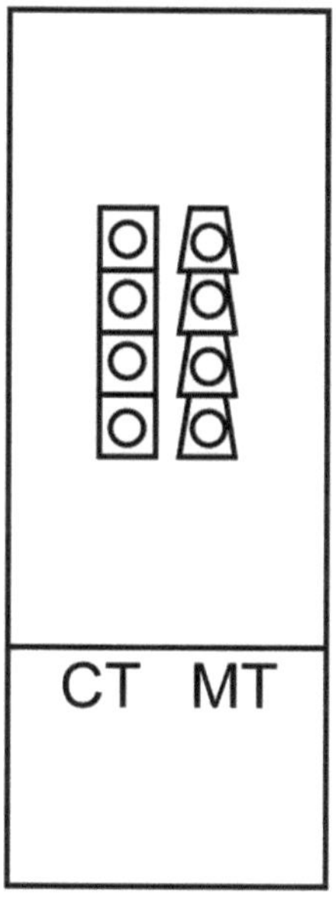

Fig. 1 Schematic illustration of a slide containing control and mutant FFPE sections. Each section is trimmed with different shapes. More consistent and comparable results can be obtained because the entire procedure is performed on the same slide

3.3 Sectioning

1. Trim a paraffin block with a razor blade so that the block is slightly larger than the sample. The top and bottom trimmed lines shall be parallel. Section the sample in 9 μm slices (*see* **Note 5**). Make paraffin ribbons containing four sections and float the ribbons on the surface of a DEPC-DW water bath at 40 °C for 1–2 min. Dip a slide into the water, scoop a paraffin ribbon onto the slide, and place the paraffin ribbon in the middle of the slide (Fig. 1) (*see* **Note 6**). Dry the slide in a slide warmer overnight. Sections are stored at room temperature. Use sections within 3 months for ISH experiments.

3.4 Pretreatment

1. Bake slide at 60 °C in a drying oven for 2 h. Baked slides can be stored at room temperature for no more than 1 week.
2. Incubate slide room temperature in xylene bath for 5 min to remove paraffin. Change xylene and repeat the incubation.
3. Wash slide in 100% EtOH for 5 min at room temperature to remove xylene. Repeat EtOH wash once.
4. Air-dry slides at room temperature for 10 min.
5. Incubate slides in 3% hydrogen peroxide at room temperature for 10 min.
6. Wash slides with DEPC-DW. Gently shake three to five times and discard DEPC-DW. Repeat once.

7. During hydrogen peroxide treatment, start to boil 100 mL of target retrieval solution in a 150 mL beaker (*see* **Note 7**). Place slides leaning on beaker wall with section side facing out (*see* **Note 8**). Reduce heat to keep target retrieval solution gently simmering for 15 min (*see* **Note 9**).
8. Immediately after boiling, wash slides in a slide container with DEPC-DW (*see* **Note 10**). Gently shake three to five times and discard DEPC-DW. Repeat DEPC-DW wash once again.
9. Wash slides with 100% EtOH twice as done with DEPC-DW in **step 8**.
10. Air-dry slides for 10–15 min at room temperature.
11. Draw a circle around sections with ImmEdge hydrophobic PAP pen two to three times to create a barrier. Minimize the area of the encircled regions (*see* **Note 11**).
12. Dry slides at 4 °C overnight.

3.5 Preparation for Hybridization and Detection

1. Warm up systems and reagents
   HybEZ oven: Warm up HybEZ oven to 40 °C. Place two pieces of threefold paper towels in a humidity control tray and wet with DEPC-DW. Close the tray lid, and place it in the HybEZ oven at 40 °C for 30 min.
   Target probes: Warm up probes in HybEZ oven at 40 °C for 10 min to dissolve any residual precipitation. Take probes out of HybEZ oven, and place at room temperature (*see* **Note 12**).
   Slides: Take slide out of 4 °C, and place at room temperature for 10 min.
   RNAscope 2.5 HD Detection Reagent: Place Amp 1–6 at room temperature 1 h before use (except Brown-A and Brown-B).
2. Prepare 1× wash buffer, 50% hematoxylin, and 0.02% ammonium hydroxide solutions.

3.6 Proteinase Treatment

1. Apply proteinase III onto sections (*see* **Notes 13** and **14**). Incubate slides in HybEZ oven at 40 °C for 30 min (*see* **Note 15**).
2. Wash slides in slide container with DEPC-DW. Gently shake three to five times and discard DEPC-DW. Repeat DEPC-DW wash once again.

3.7 Hybridization and Detection

1. Remove excess liquid from sections by flicking slide. Apply probe on sections (two drops of probe should be sufficient to cover a 2 × 2 cm encircled section) (*see* **Note 16**).
2. Incubate sections and probe at 40 °C for 2 h (*see* **Note 17**).

3. Wash sections with 1× wash buffer. Prepare 1× wash buffer in slide container close to HybEZ oven. Take the tray out from the oven. Move the slides to a slide container with sufficient 1× wash buffer. Make sure all sections are submerged in wash buffer. Close the tray, and bring it back the HybEZ oven. Gently shake the slide container two to three times, and keep it at room temperature for 2 min. Discard the wash buffer, and replace with new 1× wash buffer. Wash for another 2 min.
4. Remove excess liquid from sections by flicking the slide. Apply two drops of Amp1 on the sections one slide at a time.
5. Incubate sections with Amp1 at 40 °C for 30 min.
6. Repeat **step 3**.
7. Remove excess liquid from the sections by flicking the slides. Apply two drops of Amp2 onto sections.
8. Incubate sections with Amp2 at 40 °C for 15 min.
9. Repeat **step 3**.
10. Remove excess liquid from sections by flicking slide. Apply two drops of Amp3 onto sections.
11. Incubate sections with Amp3 at 40 °C for 30 min.
12. Repeat **step 3**.
13. Remove excess liquid from sections by flicking slide. Apply two drops of Amp4 onto sections.
14. Incubate sections with Amp4 at 40 °C for 15 min.
15. Repeat **step 3**.
16. Prepare moist staining tray at room temperature.
17. Remove excess liquid from sections by flicking the slides. Apply two drops of Amp5 onto sections.
18. Cover the staining tray. Incubate sections with Amp5 at room temperature for 30 min (*see* **Note 18**).
19. Repeat **step 3**.
20. Remove excess liquid from sections by flicking slide. Apply two drops of Amp6 onto sections.
21. Cover the staining tray. Incubate sections with Amp6 at room temperature for 15 min.
22. Remove Brown-A and Brown-B from storage to room temperature. Mix equal volumes of Brown-A and Brown-B.
23. Remove excess liquid from sections by flicking slide. Apply ~50 μL of Brown-A and Brown-B mixture onto sections.
24. Cover the staining tray. Incubate sections with Brown-A and Brown-B mixture at room temperature for 10 min.
25. Wash sections with DW. Shake gently three to five times and drain water. Wash with DW once again.

3.8 Counter Staining and Mounting

1. Place the slides in a jar with 50% hematoxylin for 1 min 15 s at room temperature (*see* **Note 19**).
2. Rinse sections briefly with DW twice.
3. Place slides in 0.02% ammonium hydroxide solution for 10 s.
4. Repeat **step 2**.
5. Dehydrate section through 30%, 50%, 70%, and 100% of EtOH at room temperature for 2 min each step.
6. Place slides in xylene at room temperature 5 min ×2.
7. Mount with Cytoseal 60.

4 Results

With RNAscope 2.5 HD Detection Reagent-BROWN, the signal is observed as brown dots. Example images using the RNAscope 2.5 HD Detection Reagent-BROWN assay are shown in Fig. 2.

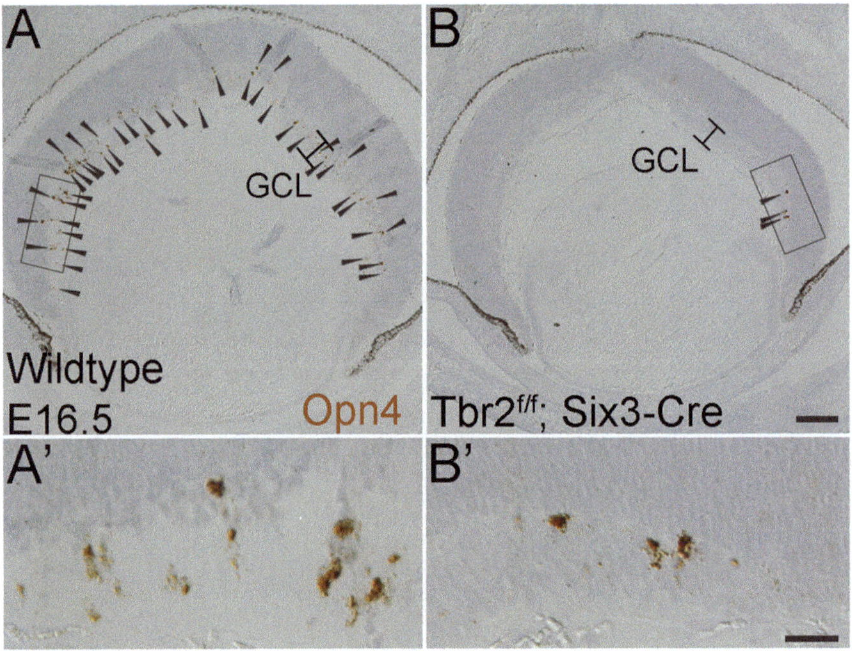

Fig. 2 Example of RNAscope brown-color reaction using Opn4 probe with wild-type (**a**) and Tbr2 mutant (**b**) embryonic eyes at E16.5. ISH signals are detected as brown dots. Opn4 is expressed in photosensitive RGCs [16]. Tbr2 regulates *Opn4* expression positively [17]. Opn4 ISH signals are detected in the GCL of wild-type retina. In contrast, Opn4 expression is significantly reduced in the Tbr2-conditional knockout in the central GCL ($Tbr2^{f/f}$; Six3-Cre, **b**). GCL are indicated with lines. Opn4-positive cells in Tbr2-mutant are indicated with arrowheads. (**a′**, **b′**) Higher magnification of Opn4 signals is indicated in areas in (**a**) and (**b**). (**c**) RNAscope brown-color reaction using Pgam2 probe with P21 wild-type retina. Pgam2 is an enzyme to catalyze a glycolysis pathway. Scattering ISH signals are detected in ONL. Weak signals are in INL and GCL. Pgam2 ISH signals in GCL are indicated with arrowheads. Scale bars: 100 μm in **b** and 25 μm in **b′**

5 Notes

1. The RNAscope assay applies novel ISH probes. A mixture of multiple probes to recognize specific target genes is designed and purchased from Advanced Cell Diagnostics (Newark, CA). These probes contain unique sequences for binding to target RNA, a spacer, and a tail sequence. A pair of tail sequences are used for signal amplification (double Z design). The double Z design enables significant target specificity and background reduction. Detailed mechanisms can be found at manufacturer's website (https://acdbio.com/science/how-it-works).
2. RNAscope ISH is compatible with flat-mount retinas. We performed RNAscope brown-color ISH with a RNAscope Jam2 probe on a flat-mounted mouse retina. Jam2, junctional adhesion molecule 2, is expressed in a subset of OFF retinal ganglion cells (RGCs) [14, 15]. Using RNAscope technology, we detected clear signals on the ganglion cell layer of the retina (Fig. 3). However, the retinas tend to disintegrate during the procedure due to the proteinase treatment.
3. Insufficient fixation will cause problems with hybridization and signal detection. Sample size should be smaller than $4 \times 4 \times 4$ mm. For mouse embryos older than E16.5, trim

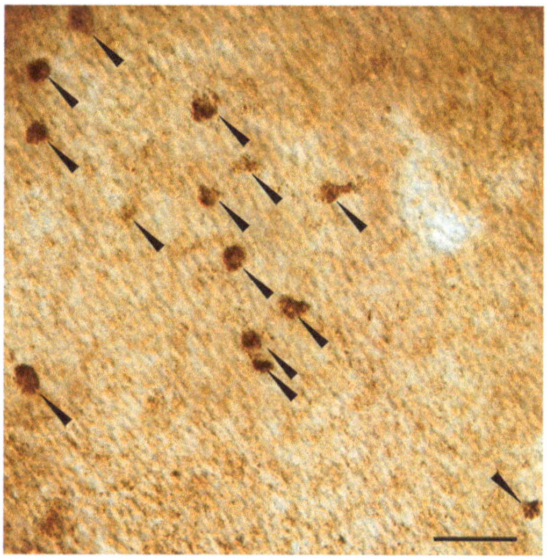

Fig. 3 Example of RNAscope brown-color ISH on flat-mounted mouse retina using RNAscope Jam2 probe. A P7 wild-type mouse retina was flat-mounted with GCL side up. For pretreatment, retinas were placed in a metal tray containing 50 mL of target retrieval solution. Then the tray was heated in a steamer for 5 min. After pretreatment, retinas were treated in a 24-well culture plate throughout the procedure. Jam2-positive RGCs are indicated by arrowheads. Scale bars: 50 μm. *GCL* ganglion cell layer

off the hind brain and nose, remove the skin covering the eyes, and poke a few holes in the head and brain with needles for better fixative penetration. For embryos between E18.5 and P0, in addition to trimming off the nose and hind brain, cut the head longitudinally in half. Dissect the eyeballs from mice older than P1. Cardiac perfusion with fixatives is necessary for larger tissues (e.g., brain) followed by longer fixation for better penetration.

4. Take sufficient time for complete penetration of EtOH during each incubation. For embryonic heads (<E15.5) or adult eyeballs, change EtOH every 1.5 h. For embryonic heads at E16.5 or older, change EtOH every 2.5 h.
5. The adult lens is highly crystalized in the paraffin block. The lens will often be cracked into small pieces and spread over the retina during sectioning. To avoid cracking the lens, briefly moisten the paraffin block with warm water at each sectioning step. The lens can be carefully removed from the slide under a dissection microscope with forceps before the section is completely dry.
6. If using two samples (e.g., control and mutant) for comparison, they should be placed on the same slide. Trim paraffin blocks in different shapes to distinguish the samples from each other (e.g., rectangular and trapezoid). Prepare paraffin section ribbons of control sample containing four sections and place on a piece of clean paper. Subsequently, prepare the mutant section ribbons of the same thickness. Float two of the control and mutant ribbons on a DEPC-DW water bath. Move the two floating paraffin ribbons with a pencil tip or wood stick next to each other. Place both ribbons side-by-side onto a slide. The ribbons can be repositioned by pushing them with the pencil tip under a dissection microscope before completely dried (Fig. 1).
7. Target retrieval solution can be replaced by other products, such as antigen unmasking solution (Vector Laboratories).
8. Avoid extensive bubbles during boiling, as the sections may become damaged.

   Place two or three slides touching each other at the bottom. Because bubbles are produced from the bottom edge of the two slides and flow up, the sections will not be damaged by the simmering bubbles (Fig. 4). A maximum of three slides can be treated in a 150 mL beaker. If treating more than three slides, prepare additional target retrieval beakers.
9. Target retrieval solution is first heated up on heating plate at the highest power. About 10 min later, the solution should start boiling. Reduce the power and keep uniform gentle boiling. Do not boil the target retrieval solution for more than

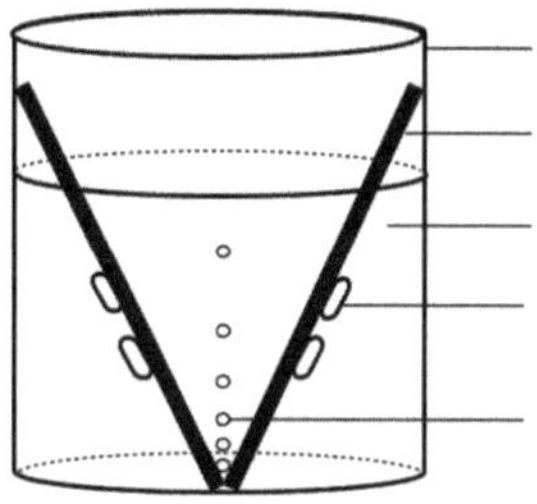

Fig. 4 Schematic illustration of the side view of the target retrieval procedure. Slides are placed in a beaker with the section side facing out so that bubbles do not disturb and damage the sections

30 min. After the tissue is placed in the target retrieval solution, monitor carefully to keep a gentle boil by controlling the heating plate power. Also, carefully monitor to ensure that all sections are submerged in the solution. If sections are above solution, reposition the slide, or add a small amount of water.

10. Do not cool slide in target retrieval solution after boiling.
11. We strongly recommend the immEdge pen. The size of the circle should hold three drops of reagents.
12. Continuous warming of the probe may cause a reduction in probe efficiency after many uses. Aliquoting and storing the probe not in use may save the probe's efficiency.
13. Proteinase IV may damage both cryo- and paraffin-embedded mouse retinal sections.
14. Because sections are dried, it is not easy to spread proteinase III on the sections. Drop a small amount of proteinase III on each section and spread with the tip of the bottle so that the proteinase III solution covers the entire section. Try not to touch the sections.
15. We have noticed that 20 min of incubation with proteinase III treatment on adult retinas gives an equally good signal as a 30 min treatment.
16. It is very important to remove excess reagent from the sample when moving on to the next treatment. On the other hand, it is necessary to avoid drying the sections. If using multiple slides, remove liquid and apply probe one slide at a time.
17. Until hybridization is completed, the whole procedure should be performed in an RNase-free environment.
18. Ensure Amp5 and Amp6 treatments are performed at room temperature.
19. Incubation in 50% hematoxylin for more than 1 min 30 s often results in a dark counter stain. On the other hand, less than 1 min of incubation will result in too faint staining.

Acknowledgments

We acknowledge grant support from the National Institutes of Health National Eye Institute (R01EY024376 and P30EY028102) and the Hermann Eye Fund.

Conflict of Interest: The authors declare no conflicts of interest.

References

1. Angerer LM, Angerer RC (1981) Detection of poly A[+] RNA in sea urchin eggs and embryos by quantitative in situ hybridization. Nucleic Acids Res 9:2819–2840
2. Larsson LI, Hougaard DM (1993) Sensitive detection of rat gastrin mRNA by in situ hybridization with chemically biotinylated oligodeoxynucleotides: validation, quantitation, and double-staining studies. J Histochem Cytochem 41:157–163
3. Femino AM, Fay FS, Fogarty K et al (1998) Visualization of single RNA transcripts in situ. Science 280:585–590
4. Nuovo GJ (1995) In situ PCR: protocols and applications. PCR Methods Appl 4:S151–S167
5. Speel EJM, Saremaslani P, Roth J et al (1998) Improved mRNA in situ hybridization on formaldehyde-fixed and paraffin-embedded tissue using signal amplification with different haptenized tyramides. Histochem Cell Biol 110:571–577
6. Player AN, Shen LP, Kenny D et al (2001) Single-copy gene detection using branched DNA (bDNA) in situ hybridization. J Histochem Cytochem 49:603–611
7. Wang F, Flanagan J, Su N et al (2012) RNAscope: a novel in situ RNA analysis platform for formalin-fixed, paraffin-embedded tissues. J Mol Diagn 14:22–29
8. Pronin A, Levay K, Velmeshev D et al (2014) Expression of olfactory signaling genes in the eye. PLoS One 9:e96435
9. Smit-McBride Z, Oltjen SL, Radu RA et al (2015) Localization of complement factor H gene expression and protein distribution in the mouse outer retina. Mol Vis 21:110–123
10. Kiyama T, Chen CK, Wang SW et al (2018) Essential roles of mitochondrial biogenesis regulator Nrf1 in retinal development and homeostasis. Mol Neurodegener 13:56
11. Stempel AJ, Morgans CW, Stout JT et al (2014) Simultaneous visualization and cell-specific confirmation of RNA and protein in the mouse retina. Mol Vis 20:1366–1373
12. Gross-Thebing T, Paksa A, Raz E (2014) Simultaneous high-resolution detection of multiple transcripts combined with localization of protein in whole-mount embryos. BMC Biol 12:55
13. Morrison JA, McKinney MC, Kulesa PM (2017) Resolving in vivo gene expression during collective cell migration using an integrated RNAscope, immunohistochemistry and tissue clearing method. Mech Dev 148:100–106
14. Daniele LL, Adams RF, Durante DE et al (2007) Novel distribution of junctional adhesion molecule-C in the neural retina and retinal pigment epithelium. J Comp Neurol 505:166–176
15. Kim IJ, Zhang Y, Yamagata M et al (2008) Molecular identification of a retinal cell type that responds to upward motion. Nature 452:478–482
16. Hatter S, Kumar M, Park A et al (2006) Central projections of melanopsin-expressing retinal ganglion cells in the mouse. J Comp Neurol 497:326–349
17. Mao CA, Li H, Zhang Z et al (2014) T-box transcription regulator Tbr2 is essential for the formation and maintenance of Opn4/melanopsin-expressing intrinsically photosensitive retinal ganglion cells. J Neurosci 34:13083–13095

Chapter 12

Single-Cell Capture, RNA-seq, and Transcriptome Analysis from the Neural Retina

Rachayata Dharmat, Sangbae Kim, Yumei Li, and Rui Chen

Abstract

Single-cell RNA sequencing (scRNA-seq) is an emerging technology that can address the challenge of cellular heterogeneity. In the last decade, the cost per cell has been dramatically reduced, and the throughput has been increased by 10^4-fold. Like many other tissues, the retina is highly heterogeneous with an estimated of over 100 subtypes of neuronal cells. Here, we describe the current techniques to perform scRNA-seq on the adult retinal tissue including retinal dissection, retinal dissociation, assessment of cell population, cDNA synthesis, library construction, and next-generation sequencing. In addition, we introduce a workflow of scRNA-seq data analysis using open-source tools.

Key words Retinal dissection and dissociation, Single-cell RNA-seq (scRNA-seq), Single-nuclei RNA-seq (snRNA-seq), ICELL8, 10× Chromium, Poly-(A)+ transcriptome amplification, ScRNA-seq analysis

1 Introduction

Individual cells serve as the building blocks of biological systems in both form and function. Understanding the individual functional characteristics of a cell type is imperative to determine its contribution to both the functionality of a normal tissue and progression of a disease state. Standard RNA-seq approaches for quantifying molecular states depend on estimating the average expression profile from thousands of individual cells. Given the heterogeneity of cell population in a biological system, measurement of average transcriptome profile masks the uniqueness in cellular function that may be crucial for maintaining normal tissue functions and fails to reveal early changes facilitating disease progression [1]. Single-cell RNA sequencing (scRNA-seq) has thus emerged as a revolutionary tool to uncover unique cellular profiles and dissect gene expression at the fundamental cellular level in different tissues [2].

ScRNA-seq technology has been recently applied to multiple tissues and has led to the discovery of novel cell types [3]

Chai-An Mao (ed.), *Retinal Development: Methods and Protocols*, Methods in Molecular Biology, vol. 2092,
https://doi.org/10.1007/978-1-0716-0175-4_12,

elucidating trajectories of distinct cellular lineages during development [4, 5] and uncovering regulatory relationships between genes [6], among other discoveries. The retina has been successfully utilized to elucidate the intricacies of its various cell types and their organization [7]. The retina is a highly specialized and polarized layer of neurons that lines the back of the eyeball and functions in transduction of visual signals. It is an ordered array of >60 distinct types of cells spanning six major classes: photoreceptors (rods and cones), bipolar cells (BPC), amacrine cells (AC), horizontal cells (HC), retinal ganglion cells (RGCs), and Müller glial cell (MG) [8]. Hallmark studies have discovered novel retinal neuron subtypes by profiling population of amacrine, bipolar, and retinal ganglion cells by scRNA-seq [7, 9, 10]. These studies have revealed novel retinal cell types based on their gene expression and have provided unique markers to further identify and dissect these rare cells functionally. These studies thus exhibit utility of scRNA-seq and provide insights into diverse biological systems.

Since the first demonstration of RNA-seq from a single cell in 2009 [11], several scRNA-seq studies using different approaches have been reported over the past decade. The basic scRNA-seq methodology consists of four steps: (1) isolation and lysis of single cells or single nuclei, (2) reverse transcription (RT), (3) cDNA amplification, and (4) sequencing library preparation [12]. Single-cell isolation begins by dissociating the target tissue by enzymatic digestion to yield a high-quality viable cell population. These dissociated cells are then sorted or isolated into individual reaction volumes that allow downstream processing. While easier for cells lacking strong extracellular matrix, such as hematopoietic cell populations, this step requires optimization for more sensitive tissues such as the brain or the retina. Further enrichment of specific cell types using pan-cell type surface markers can be performed at this stage to allow downstream capture and processing of the target population such as bipolar cells from the retinal population [10]. The rate-limiting step of single-cell isolation has recently been improved with the use of microfluidic technology. Microfluidic devices can capture single cells or single nuclei in isolated reaction chambers, i.e., nanowells or microdroplets. The cells are then lysed to release the RNA and subjected to RT in automated nanoliter reactions [13–15]. These technologies have further been commercialized (ICELL8 (https://www.takarabio.com/products/automation-systems/smarter-icell8-systems/smarter-icell8-single-cell-system) and 10× genomics (http://go.10xgenomics.com/single-cell-controller)) to provide kits which have improved the utility of scRNA-seq from specialized technique to a generally applicable technology. Based on the underlying biological question, the choice of methods for downstream RT and cDNA amplification in scRNA-seq can vary. For most purposes of quantifying gene expression profile, poly(A)+ selection is performed using poly (dT) primers to capture messenger RNAs (mRNAs). Next, utilizing the "switching mechanism at 5′ end of RNA template" (SMARTer) chemistry for mRNA, reverse transcription and cDNA synthesis are performed. The small amount of cDNAs generated is then further amplified using conventional PCR. Amplified and bar-coded cDNA (that shares the identity of the cell as bead bar code or well bar code in 10× chromium or ICELL8, respectively) from every cell is pooled and sequenced by NGS, using Nextera-based library preparation techniques, and subjected to downstream sequencing.

In this protocol, we highlight the techniques to perform scRNA-seq on the retinal tissue. The protocol begins with steps that involve retinal dissection, retinal dissociation, and assessment of cell population. Since fresh retina can only be obtained in limited scenarios such as mouse models, this protocol also includes the alternative approach of retinal single-nuclear RNA-seq (snRNA-seq). Retinal cell nuclei can be extracted from either fresh or snap-frozen retinal specimen and can be subjected to the same downstream processing as live cells. Two alternative and most readily accessible commercial methods for single-cell capture are highlighted in this protocol: nanowell microfluidics-based ICELL8 and microdroplet chemistry (drop-seq)-based 10× chromium.

In parallel with the improvements in methods of scRNA sequencing technologies, many downstream data analyzing tools have been developed [16]. A set of public open resource tools can be found at the scRNA-tools database (https://www.scRNA-tools.org), Bioconductor (https:www.bioconductor.org), PyPI (https://pypi.org), and GitHub (https://github.com). However, many challenges still remain in scRNA-seq data analysis due to factors such as low sensitivity, data sparsity, batch effects, and high complexity of the data [17].

Here, we provide an example procedure for analyzing scRNA-seq data starting from read count or the unique molecular identifier (UMI) profile matrix to decipher the biological significance, using the Seurat package (https://satijalab.org/seurat/), one of the most commonly used tools. In brief, the data processing step includes quality control (QC), classifying cell population, identifying cell type-specific genes, and pseudo-time analysis. Different workflow and tools can be used depending on the study design and biological questions. Briefly, after mapping the read sequences to the reference genome using alignment software, such STAR, HISAT, and Bowtie, the read count or UMI matrix with genes in row and cells in column can be generated [18–20]. Data QC is performed to filter low-quality cells and genes by the proportion of unexpressed genes, followed by normalization and scaling of the filtered matrix data. Next, the high-dimensional data are visualized using PCA, canonical correlation analysis (CCA), or t-Distributed Stochastic Neighbor Embedding (t-SNE, https://lvdmaaten.github.io/tsne/)

to investigate cell subpopulation structures. The identity of each cluster is determined based on the expression pattern of highly variable genes or marker genes. Significantly expressed genes unique to each cell cluster can be identified based on comparing with the rest of the cluster. In addition, Gene Ontology (GO) or pathway analysis of the significant genes can be performed for differentially regulated signaling pathways in each cluster.

2 Materials and Equipment

2.1 Equipment

2.1.1 General Equipment

1. Thermal cycler for 0.2 mL tubes
2. Bioanalyzer instrument (Agilent)
3. Qubit Fluorometer (Thermo Fisher Scientific)
4. PCR hood system (UV equipped)
5. DNA LoBind tubes, 2.0 mL

2.1.2 Retinal Dissection and Dissociation

1. Dissection scope
2. Fine microdissection scissors (Vannas Spring Scissors, 2 mm cutting edge)
3. Forceps (Student Dumont #5 and #2)
4. Precision™ general purpose water baths
5. Round-bottom tubes
6. Glass Dounce homogenizer (1 mL, 7 mL)
7. 20 μm Sieve cloth or Falcon Cell Strainers (mesh size, 40 μm)
8. 0.22 μm Syringe filter and 10 mL syringe

2.1.3 ICELL8 System

1. ICELL8 MultiSample NanoDispenser (MSND; WaferGen 640000)
2. ICELL8 Imager, ICELL8 Imager Microscopy suite
3. CellSelect image analysis software
4. ICELL8 chip holder
5. Centrifuge chip spinner, chip cycler (T-100) with chip adapter
6. Film applicator
7. Chip cold block and MSND 384 well plates and seals

2.1.4 10× Chromium System

1. Chromium™ Single Cell Controller (120263)
2. Vortex adapter (10× chromium, 120251)
3. Chip holder (10× chromium, 120252)

2.2 Reagents and Stock Solutions

2.2.1 Retinal Dissociation

1. MEM + 20%FBS: 40 mL MEM w/o glutamine + 10 mL heat-inactivated FBS/FCS.
2. 10× Low ovomucoid (LO) solution: 150 mg BSA + 150 mg ovomucoid + 10 mL MEM w/o glutamine. Dissolve and adjust the pH to 7.4. Filter through a 0.22 μm syringe filter, and freeze in 1 mL aliquots.
3. 160,000 U/mg DNase I: working stock, 80 U/μL reconstituted in 10 mM Tris–Cl (pH 7.5) + 150 mM NaCl+2 mM $MgCl_2$ solution. Filter through a 0.22 μm syringe filter, and freeze in 100 μL aliquots.
4. L-Cysteine stock: 40 mg L-cysteine, anhydrous + 1 mL ddH_2O. Filter through a 0.22 μm syringe filter, and freeze in 100 μL aliquots.
5. Other reagents: HBSS (with phenol red indicator) and PBS (−/−), water (molecular biology grade, D-RNase-free).

2.2.2 ICELL8 Reagents

1. ICELL8 Single-Cell Poly-(A)+ transcriptome amplification reagent kit (WaferGen 640142)
2. 1×PBS (no Ca^{2+}, Mg^{2+}, phenol red, or serum; pH 7.4)
3. ReadyProbes® Cell Viability Imaging Kit, blue/red (contains Hoechst 33342 and propidium iodide)
4. Murine RNase inhibitor (40,000 U/mL)
5. Agencourt AMPure XP magnetic beads
6. High-sensitivity DNA analysis kit for Bioanalyzer (110 samples; Agilent: 50674626)
7. Nextera XT DNA Library Preparation Kit (24 samples; Illumina: FC-131-1024)
8. Nextera XT Index Kit (24 indexes, 96 samples; Illumina: FC-131-1001)
9. Water (D-RNase-free, molecular biology grade)
10. ICELL8 RT-PCR mix: 56 μL betaine (5 M)* + 24 μL of 25 mM dNTP mix* + 3.2 μL $MgCl_2$ (1 M)* + 8.8 μL DTT (100 mM)* + 61.9 μL SMARTScribe™ First-Strand Buffer (5×)* + 33.3 μL SeqAmp™ PCR Buffer (2×)* + 8.8 μL RT E5-oligo (100 μM)* 4.0 Amp Primer (10 μM)* + 1.6 μL Triton X-100 (100%) + 28.8 μL SMARTScribe™ Reverse Transcriptase** + 9.6 μL SeqAmp™ DNA Polymerase**

    *Mix these components by vortexing. Centrifuge to collect. Add the Triton X-100. Vortex to mix until the Triton is dissolved.

    **Add the enzymes, and mix by pipetting up and down.

2.2.3 10× System Reagents

1. Chromium Single Cell 3′ Library Kit v2: 4 reactions (chromium 120264)
2. Chromium Single Cell 3′ Gel Bead Kit v2 (120265)
3. Dynabeads MyOne SILANE: 16 reactions (2000048)
4. Chromium Single Cell A Chip Kit: 16 reactions (1000009)
5. Chromium Multiplex Kit: 96 reactions (120262), sample index plate (10× genomics 220103)

3 Methods

3.1 Retinal Dissection

1. Carefully remove the eyeballs from a euthanized mouse, and collect them in a 35 mm dissection dish containing 3 mL D-PBS (+/+) (*see* **Note 1**).
2. Position the eye cup in place by holding down the optic nerve using a Student Dumont #5 forceps. Make a small puncture using a 22-G needle at the center of the cornea.
3. Using a microdissection scissor, make four incisions (in an "X" shape) in the cornea with the hole as the center. The cuts must be deep enough to reach the corneal–scleral divide. Hold the eye cup at the optic nerve to position, and use the #5 forceps to open the X-shaped incision and remove the lens.
4. Using the #2 and #5 forceps, gently pull at the corneal incision so as to release the retinal cup. Use the #2 forceps to position the scleral cup, and gently push at the neural retina to detach it from the RPE layer. Using #5 forceps, remove any residual RPE attached. Finally, make a cut at the base of the optic nerve from the internal attachment with angled scissors (*see* **Note 2**).
5. The retinal cup is ready for dissociation (Fig. 1a).

3.1.1 Preparation of Single-Cell Capture Platforms

Critical: All steps require meticulous execution to avoid contamination. It is important to work in a PCR hood whenever possible. And wear clean, protective laboratory coat in the hood prior to amplification. All reagents and consumables must be free of DNA, RNA, and nucleases. The quality of a single-cell reaction can be significantly affected by small changes in reagents (concentration, age, or manufacturer). Avoid multiple freeze-thaw cycles (aliquot reagents). Prior to beginning, clean all surfaces to remove nucleic acid and nuclease contamination.

3.1.2 Retina Dissociation Procedure (Fig. 1c)

1. Transfer to a round-bottom tubes (Falcon 2059) containing HBSS (pH 7.4, with phenol red indicator).
2. While the retina is being dissected, make papain-based dissociation solution by mixing 5 mL of HBSS (with phenol red indicator), 45 U of papain (35 μL of 1.17 U/μL), and 30 μL

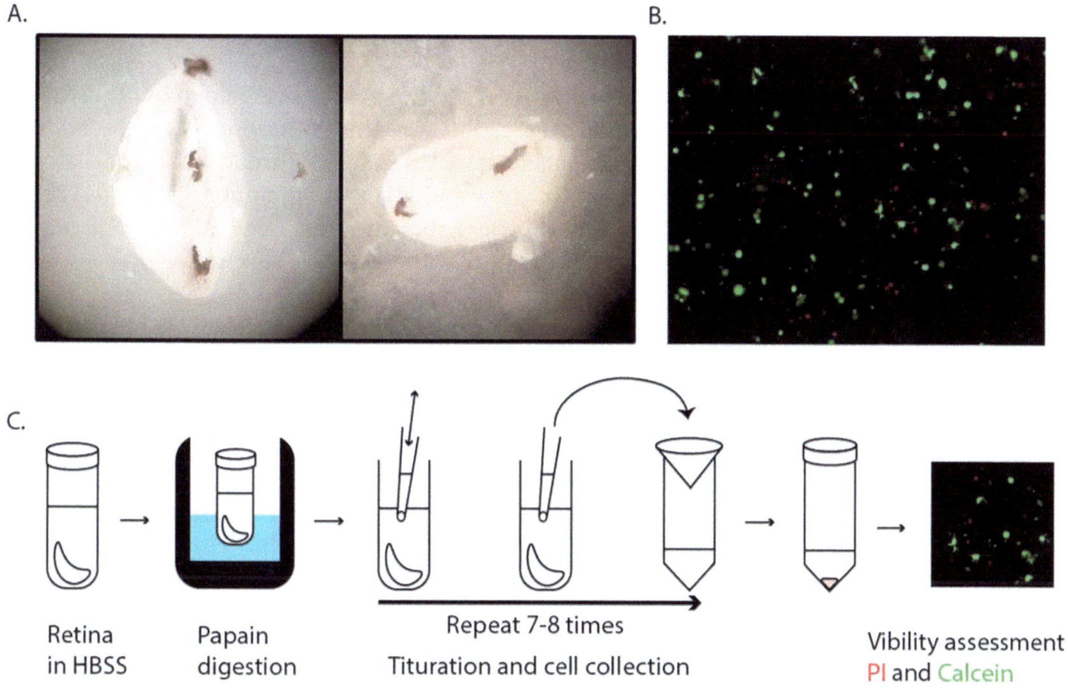

Fig. 1 Retinal dissection and dissociation: (**a**) intact retinal cups after dissection; (**b**) dissociated retina cells stained with propidium iodide (red) and calcein, AM (green) display isolated live retinal neurons; (**c**) outline of the retina dissociation procedure

L-cysteine stock (1.2 mg/prep). Add 1 μL of 2 N NaOH if there is a strong change in pH indicator of HBSS to yellow. Activate at 37 °C for 10–15 min (*see* **Note 3**). Filter the solution through a 0.22 μm syringe filter. Finally, add 15 μL of DNase I stock to the dissociation solution (1200 U/prep). The papain dissociation solution is ready for retinal dissociation (*see* **Note 4**).

3. Gently remove HBSS solution from the round-bottom tubes containing the retina, sparing approximately 200 μL to leave the retina cup undisturbed. Slowly add the 5 mL activated papain dissociation solution.
4. Incubate retinas in papain for 15–20 min, gently swirling every 4 min (*see* **Note 5**).
5. During the incubation period, make the 1× low ovomucoid (LO) solution by mixing 9 mL of MEM with 1 mL of 10× low ovomucoid solution and 15 μL of DNase I stock (1200 U/prep). Filter the solution through a 0.22 μm syringe filter.
6. After incubation, gently replace papain with 2 mL LO. The supernatant does not contain cells and can thus be discarded. Add another 2 mL of LO, and gently triturate the solution with p1000 tip 2× avoiding air bubbles. At this stage, do not

aspirate the retina during pipetting. Spin the tubes at 200 × *g* for 20 s.

7. Place a 20 μm sieve on the rim of a 50 mL Falcon by adjusting it into a funnel shape. Pre-wet this filtration system by adding 0.5 mL heat-inactivated FBS and 0.5 mL LO to the Falcon. Transfer 1 mL of the supernatant (containing dissociated cells) from the dissociation tube into the new 50 mL Falcon by straining them through the pre-wet 20 μm sieve.
8. To the remaining supernatant and tissue in the dissociation tube, add 1 mL LO solution, and triturate gently with P1000 2–5×, spin 20 s on speed 3, add supernatant to the 50 mL Falcon tube, and repeat the step until all/most cells are dissociated.
9. For final dissociation, when remaining pellet consists of a small amount of tissue, aspirate off LO and resuspend cells more vigorously with a pipette. Performing this trituration correctly and gently is critical to get a high yield of live cells.
10. Spin the collected cells at 375 × *g* at 4 °C for 10 min. Gently resuspend the pellet containing cells in 500μL of MEM/-MEM + 20%FBS. Confirm cell viability using trypan blue counterstain or viability dyes such as Hoechst 33342 and propidium iodide (dead cell stain). The yield from two adult mouse retina ranges from 6 to 6.5 × 10^6 cells/mL (Fig. 1b).

3.1.3 Retina Nuclear Isolation

1. Transfer the dissected retina from the PBS solution into a 1.7 mL Eppendorf tube using a cut P1000 tip. Remove the extra PBS solution, and wash the retina cup in the nuclear dissociation buffer. Again, remove the extra solution above the retinal cup, and add 700 μL of nuclear dissociation buffer to the retina.
2. Using a P1000 pipette set to 500 μL, triturate the tissue in the nuclear dissociation buffer to break the eye cup.
3. Transfer the nuclear dissociation buffer containing the disintegrated retinal tissue to a prechilled 2 mL Dounce tissue homogenizer placed in ice. Using the loose pestle homogenize (15–20 strokes) till tissue particles are not visible.
4. Replace the loose pestle with the prechilled tight pestle, and homogenize by 15 strokes. Incubate in ice for 1 min, and repeat homogenization with 15 strokes. Avoid air bubbles (*see* **Note 6**).
5. Collect the homogenized 700 μL of nuclear solution in a LoBind 2 mL Eppendorf tube. Wash the Dounce homogenizer and tight pestle with an additional 700 μL of nuclear dissociation buffer, and add this solution to the same Eppendorf containing the 700 μL of lysed retinal nuclei solution. Centrifuge the nuclei solution at 500 × *g* for 10 min at 4 °C.

6. Discard the supernatant, and resuspend the pellet in 1 mL of nuclear dissociation buffer using the p1000 tip.
7. Transfer the solution to the prechilled Dounce tissue homogenizer. Using the tight pestle, homogenize the nuclei 10 times. Incubate in ice for 1 minute and repeat the homogenization 10 times.
8. The solution is ready for counting assessment using a counterstain like trypan blue.

3.2 Single-Cell Capture Platform

3.2.1 The ICELL8 Single-Cell Poly-(A)+ Transcriptome Amplification

1. Stain the cell suspension (MEM) using Hoechst 33342 and propidium iodide (dead cell stain), and incubate for 5 min (the viability must be >95%). Use a 100 μL aliquot to dilute the cell suspension to a concentration of one cell/5 nL using 1×PBS (Ca-/Mg-free, pH 7.4). Do not vortex (*see* **Note 7**).
2. Spin one million cells at 375 × *g* at 4 °C for 10 min, and resuspend the pellet in TRIzol. Incubate the TRIzol at room temperature for 10 min with intermittent vortexing to resuspend any residual clumps. Store at −80 °C. This will serve as a bulk RNA control.

   Important: Instructions below are an extension of the ICELL8 Single-Cell Poly-(A)+ transcriptome amplification reagent kit.
3. Prepare the stained cell dispensing mix by mixing 10 μL of cell diluent, 10 μL of RNasein, and 100 μL of stained cell suspension of + 880 μL of 1×PBS (Ca-/Mg-free, pH 7.4). Prepare the positive control, negative control, and fiducial mix according to the manufacturer's suggested guidelines.
4. In a 365 well plate, carefully add 80 μL of cell stained cell dispensing mix in wells A1, B1, C1, D1, A2, B2, C2, and D2 by reverse pipetting method to avoid bubbles. Similarly, add 25 μL of negative control to well A24, 25 μL of positive control to well P24, and 25 μL of fiducial mix to well P1.
5. Use the ICELL8 MultiSample NanoDispenser (MSND) from the SMARTer ICELL8 Single-Cell System set to "cell dispense" setting to dispense stained live cells into a SMARTer ICELL8 chip (Fig. 2a) (*see* **Note 8**).
6. Following dispense, seal the chip with the imaging film provided in the kit. Centrifuge at 300 × *g* for 5 min at room temperature.
7. Place the chip on the microscope holding platform, and peel off the top liner using tweezers. Use the stage position list in the ICELL8 Imager Microscopy suite to set the stage at "Pos0," and illuminate the DAPI channel. Once optimal contrast is reached, focus on single cells in the center of the well. Check fiducial cells and alignment, and proceed with automatic fluorescent imaging of the chip.

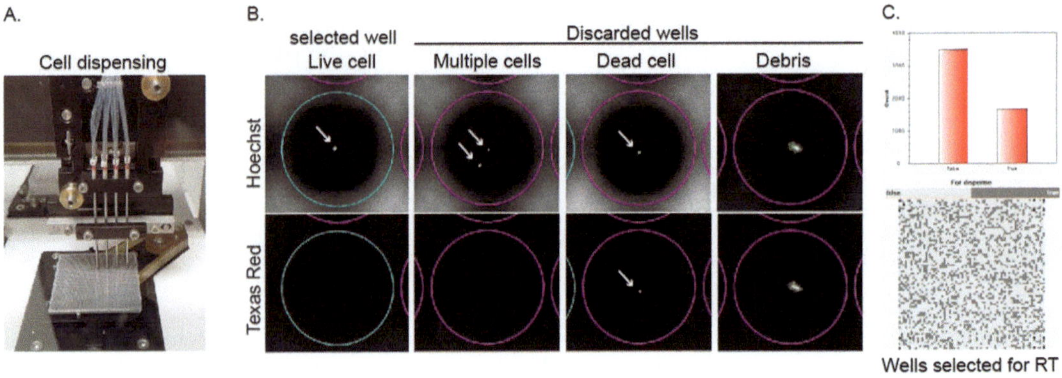

Fig. 2 (**a**) ICELL8 MultiSample NanoDispenser (MSND) system that can dispense nanoliter volume of fluid to deposit single cell per well in the ICELL8 chip. (**b**) Candidate wells selected for downstream RT have a single live cell per well (arrow). Wells containing multiple cells, dead cells (signal in the Texas Red channel), or debris are discarded from the candidate list. (**c**) CellSelect software generates a map (bottom panel) for candidate wells (approximately 1500) which are quantified in the top panel

8. Once image acquisition is completed, reapply the peeled liner on the top side of the double-sided film. Place the chip into a prechilled (−80 °C) chip holder, and transfer it to −80 °C for a minimum of 30 min. The protocol can be paused here until further processing.
9. Analyze images with CellSelect™ by loading the image files and bar code information and finally selecting "process images." This will automatically identify and select all chip wells that contain a single live cell as well as control wells. The signal in Texas Red channel will be absent for live cells. Manually check the candidate wells (approximately 1500) to ensure absence of doublets and lack of debris in the candidate wells (Fig. 2b, c). For single nuclei capture, only the DAPI channel will display a positive signal. Manual assessment of candidate wells is necessary with nuclear capture to ensure lack of doublets or aggregates which are more prevalent with nuclear capture.
10. Remove the chip from −80 °C, and thaw at room temperature for 10 min to lyse cells. Meanwhile, prepare the RT reagent mix by mixing the following reagents in the order provided: 88 μL 5 × RT buffer, 44 μL 10 mM RT dNTPs, 4.4 μL 100 μM RT E5-oligo, 57.2 μL D-RNase-free water, and 26.4 μL 200 U/μL RT enzyme.
11. Load 50 μL of the RT mix in wells A1, B1, C1, and D1 of the 365 well source plate. Place the chip on the ICELL8 MSND platform chip holder, and load the dispense file on the software. Dispense the RT-PCR mix into the selected wells as determined by the dispense file by selecting "dispense RT buffer" function.

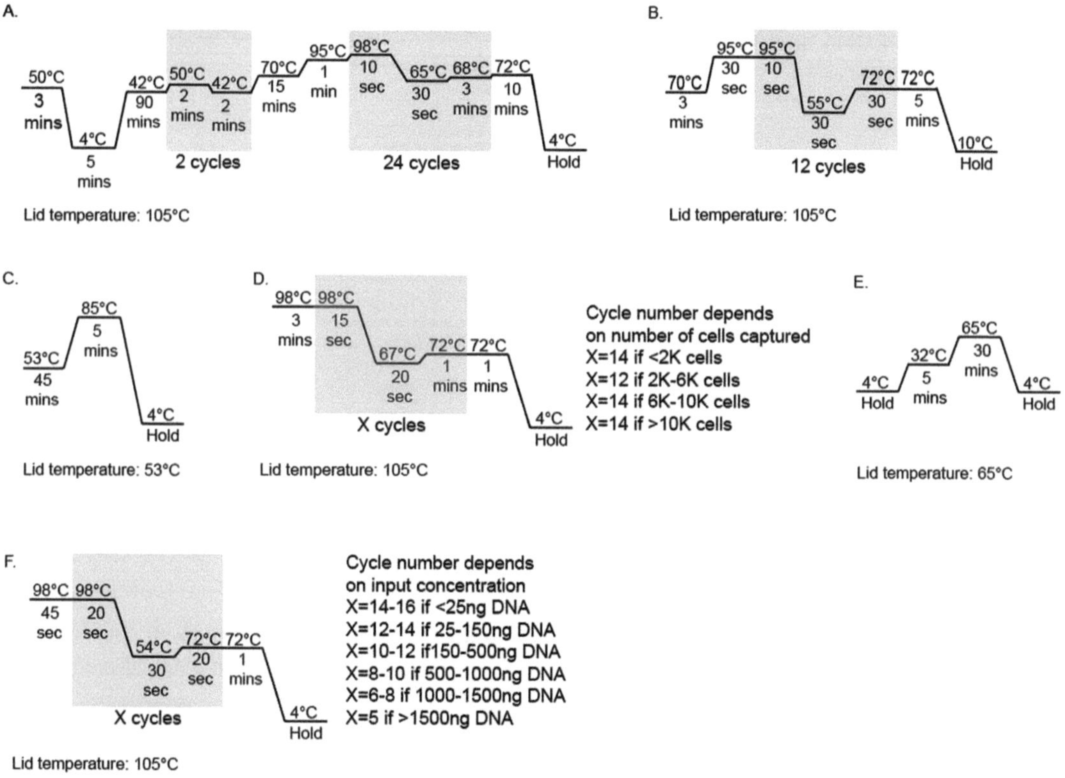

Fig. 3 Thermo-cycling parameters for ICELL8 (**a**, **b**) and 10× single cell (**c**, **f**) transcriptome amplification and library preparation

12. After the dispensing step is completed, seal the chip with a PCR film, and centrifuge at 3220 × *g* for 3 min at 4 °C. Place the chip in the SmartChip™ cycler and thermal cycling program at the parameters specified in Fig. 3a.
13. Centrifuge the chip at 3220 × *g* for 3 min at 4 °C.
14. Assemble the collection module (with the chip facing down), and seal the chip using the provided collection film. Centrifuge at 3220 × *g* for 10 min at 4 °C. Based on the number of wells selected for dispense, the volume can range from 250 to 500 μL.
15. Concentrate the cDNA product using a Zymo Clean & Concentrator kit using seven volumes of DNA binding buffer. Elute the product twice using 20 and 10 μL RNase-free water.
16. Next, purify the cDNA with 0.6× AMPure XP beads (e.g., to 30 μL of amplified cDNA product, add 18 μL of beads). Elute the reaction with 12 μL of water (*see* **Note 9**).
17. Examine quality of the cDNA product with the Bioanalyzer (Fig. 4a). Quantify the cDNA with the Qubit® dsDNA HS fluorometric assay, and use 0.2 ng/μL for library preparation.

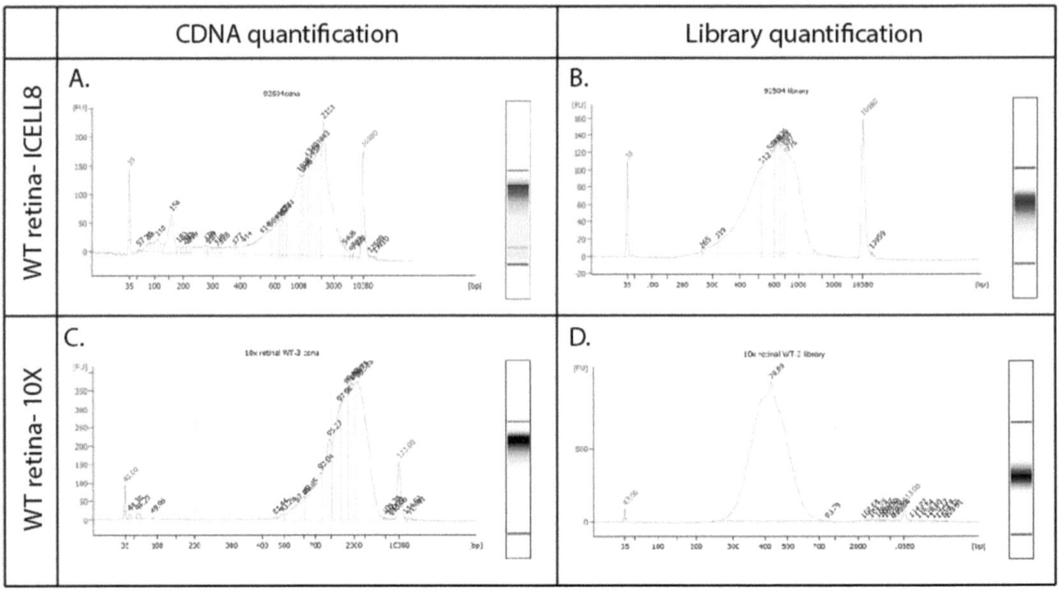

Fig. 4 Bioanalyzer electrophoresis displaying the quality of the CDNA (**a**, **c**) and library (**b**, **d**) obtained from ICELL8 Poly-(A)+ transcriptome amplification (**a**, **b**) and 10× chromium 3′ transcriptome amplification (**c**, **d**) from mouse retina cells

18. Prepare the Nextera fragmentation reaction using 10 μL of TD buffer, 5 μL of CDNA, and 5 μL of amplicon tagment mix from the Nextera XT tagmentation kit. Load the samples on a thermal cycler, and run for 55 °C for 5 min. Neutralize the reaction by adding 5 μL of neutralize tagment (NT) buffer, and pipette five times to mix the reaction. Incubate at room temperature for 5 min.
19. Prepare the PCR reaction by adding 15 μL of Nextera PCR master mix, 5 μL of i7 index primers, and 5 μL of P5 primers to 25 μL of tagmented cDNA mix. Run the reaction in a thermal cycler using parameters specified in Fig. 3b.
20. Purify the library using a 1:1 ratio followed by 0.5:1 ratio AMPure XP beads. Save the supernatant from the second reaction, and treat it with 0.2:1 ratio of beads. Elute using 11 μL nuclease-free water to obtain >15 nM of final library.
21. Examine the library quality with the Bioanalyzer (Fig. 4b), and quantify using KAPA Quant. The library is sequenced at the desired depth based on the number of candidate cells required.

3.2.2 3′ Transcriptome Amplification Using Chromium™ Single Cell 3′ v2 Kit

1. Thaw the 3′ Single-Cell Gel Beads 30 min before use. Equilibrate RT reagent mix, RT primer, and additive A to room temperature. Vortex RT primer and additive A, and prepare 50% glycerol solution. Begin the equilibration step during the final steps of retinal dissociation and trituration to avoid cell death associated with increased incubation of dissociated retinal cells.

2. Prepare the RT reaction on ice by adding 50 μL of RT reagent mix, 3.8 μL of RT primer, 2.4 μL of additive A, and 10 μL of RT enzyme mix to make a master mix for one reaction. For scaling up the reaction, please refer to Chromium™ Single Cell 3′ v2 Kit.
3. Once the retinal dissociation is complete and cell count estimated, use the cell suspension volume table provided with the Single Cell 3' Reagent Kits v2 reference guidelines. This will allow the estimate of cell suspension volume required to attain the target cell capture number. For example, with an initial cell concentration of 1×10^6 cells, it is possible to capture anywhere from 2000 cells to 10,000 cells (*see* **Note 10**). For a capture target of 10,000 cells, 17.4 μL of cell suspension and 16.4 μL of nuclease-free water are used per reaction. For dissociated retina tissue, keep the cells in MEM + 10% FBS to keep the cells healthy.
4. Dispense 66.2 μL of RT master mix into a well of a prechilled (4 °C) PCR tube strip on ice. Add the nuclease-free water and cell suspension volume as determined in the previous step to the master mix.
5. Prepare the chip such that wells that will not contain the master mix (incase loading <8 samples per chip) in row 1, 2, and 3 have been loaded with 50% glycerol according to manufacturer's guidelines. Next, transfer 90 μL of master mix containing cells to the well in the row labeled 1.
6. Vortex Single-Cell 3' Gel Bead Strip for exactly 30 s, and puncture the foil seal using a pipette tip. Carefully aspirate 40 μL of the gel beads, and load it in the well in row labeled 2 without introducing bubbles. Pipette 270 μL of partitioning oil in the well in row labeled 3, and finally, seal the chip with the 10× gasket.
7. To run the chip, eject the tray from the Chromium Controller, and position the chip with the gasket on the tray. Retract the tray, and select Single Cell A program on the screen of the controller. Press the play button. In 6.5 min, the run will be completed. Immediately proceed to the next step.
8. Prechill an emulsion safe PCR tube strip. Remove the Chip A Single Cell from the tray, and remove the gasket. Set the 10× chip holder at 45° angle. With the tips positioned vertically in the well, slowly aspirate 100 μL of GEMs from the recovery wells labeled with an arrow head. Verify the absence of air bubbles in the pipette tips. Dispense the GEMs to the prechilled PCR tube.
9. GEM-RT: Load the PCR tube in a thermal cycler that can run a reaction volume of 125 μL and an adjustable lid temperature. Run the thermal cycler at parameters defined in Fig. 3c. Upon

completion, the PCR tube can be stored at 4 °C for up to 72 h or at −20 °C for up to a week.

10. Perform a post GEM cleanup using manufacturer's protocol. Prepare the cDNA amplification reaction by mixing 1 μL of nuclease-free water, 50 μL of amplification master mix, 5 μL of cDNA additive, and 2 μL of cDNA primer mix. Add the amplification reaction to the 35 μL of eluted reaction from GEM cleanup. Run the reaction in a thermal cycler using conditions specified in Fig. 3d. The optimal number of cycles for the cDNA amplification reaction should be adjusted according to the number of cells captured to minimize PCR amplification artifacts. Use 14 cycles for 2000 cells, 12 cycles for 2000–6000 cells, and 10 cycles for 6000–10,000 cells captured in the GEM generation step. The samples can be stored at 4 °C for up to 72 h.
11. Clean up the reaction using 0.6 volume SPRI select reagent using manufactures protocol. Elute the DNA in 40 μL of elution buffer. The samples can be stored at 4 °C for up to 72 h. Quantify the CDNA sample by running 1μL of 1:5 times diluted sample (Fig. 4c).
12. For Nextera library preparation, begin by running the thermal cycler to parameters specified in Fig. 3e. To a PCR strip (kept at 4 °C) containing 35 μL of cDNA, add 15 μL of fragmentation mix (10 μL of fragmentation enzyme blend and 5 μL of fragmentation buffer provided with the kit). Load the reaction in the precooled thermal cycler, and press "SKIP" to initiate the fragmentation, end repair, and A-tailing protocol.
13. Perform double-sided SPRI selection to clean up the reaction as per manufacturer's protocol. To 50 μL of elute obtained in size selection, add 50 μL of adaptor ligation mix (20 μL ligation buffer, 10 μL DNA ligase, 2.5 μL adaptor mix, 17.5 μL nuclease-free water). Mix the sample by pipetting, and run the thermal cycler at 20 °C for 15 min.
14. Perform SPRI selection (0.8×) to clean up the reaction as per manufacturer's protocol. To 30 μL of elute obtained in size selection, add 60 μL of sample index PCR mix (50 μL amplification master mix, 2 μL SI-PCR primer, and 8 μL nuclease-free water). Add 10 μL of Chromium i7 Sample Index. Mix the reaction by pipetting, and run thermal cycler reaction at the conditions specified in Fig. 3f.
15. Perform double-sided SPRI selection to clean up the reaction as per manufacturer's protocol. Run 1 μL of the library at 1:10 dilution on the Bioanalyzer (Agilent) to determine the quality of the library generated (Fig. 4d).
16. The library is sequenced at the desired depth based on the number of candidate cells required.

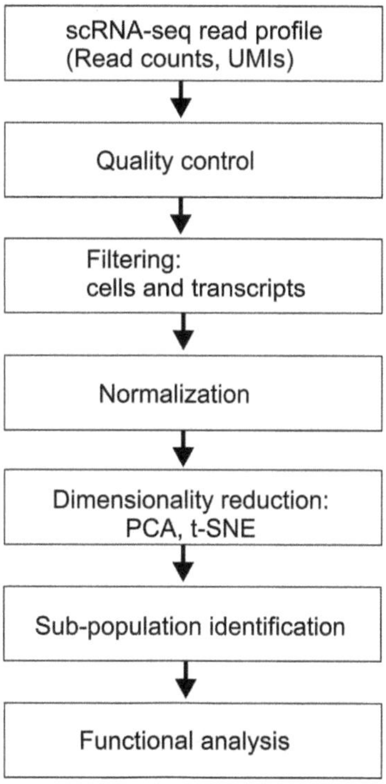

Fig. 5 Schematic view of scRNA-seq data analysis. Read counts or UMI matrix data can be processed at gene or transcript level for QC using Seurat package. Next, data normalization and multidimensional deduction can be performed, and clustered data are visualized with t-SNE plot. Each cell type is identified by the expression of known cell-specific marker genes in each cluster. Lastly, GO analysis or other pathway analysis tools can be used for characterizing the biological signature in each cluster

3.3 Single-Cell RNA-Seq Data Analysis

The scRNA-seq data analysis can be conducted by the following steps: data QC, normalization, dimensionality reduction, subpopulation identification, and functional analysis (Fig. 5). Here, we demonstrate processing the sample organoid scRNA-seq data, classifying cell subpopulations, and identifying marker genes that are differentially expressed among subclusters in retinal organoids (Fig. 6) (*see* **Note 11**).

3.3.1 Data Process

1. Download the read count profile data.

   For practice, scRNA-seq data of 8-month human retinal organoid can be downloaded from the GitHub link (https://github.com/sangbaekim/scRNAseq_MMB). The profile data were generated from aligned bam file using the HTseq-count tool (https://htseq.readthedocs.io) after aligning to hg19 reference genome using STAR. All R scripts for the practice are available at the link.

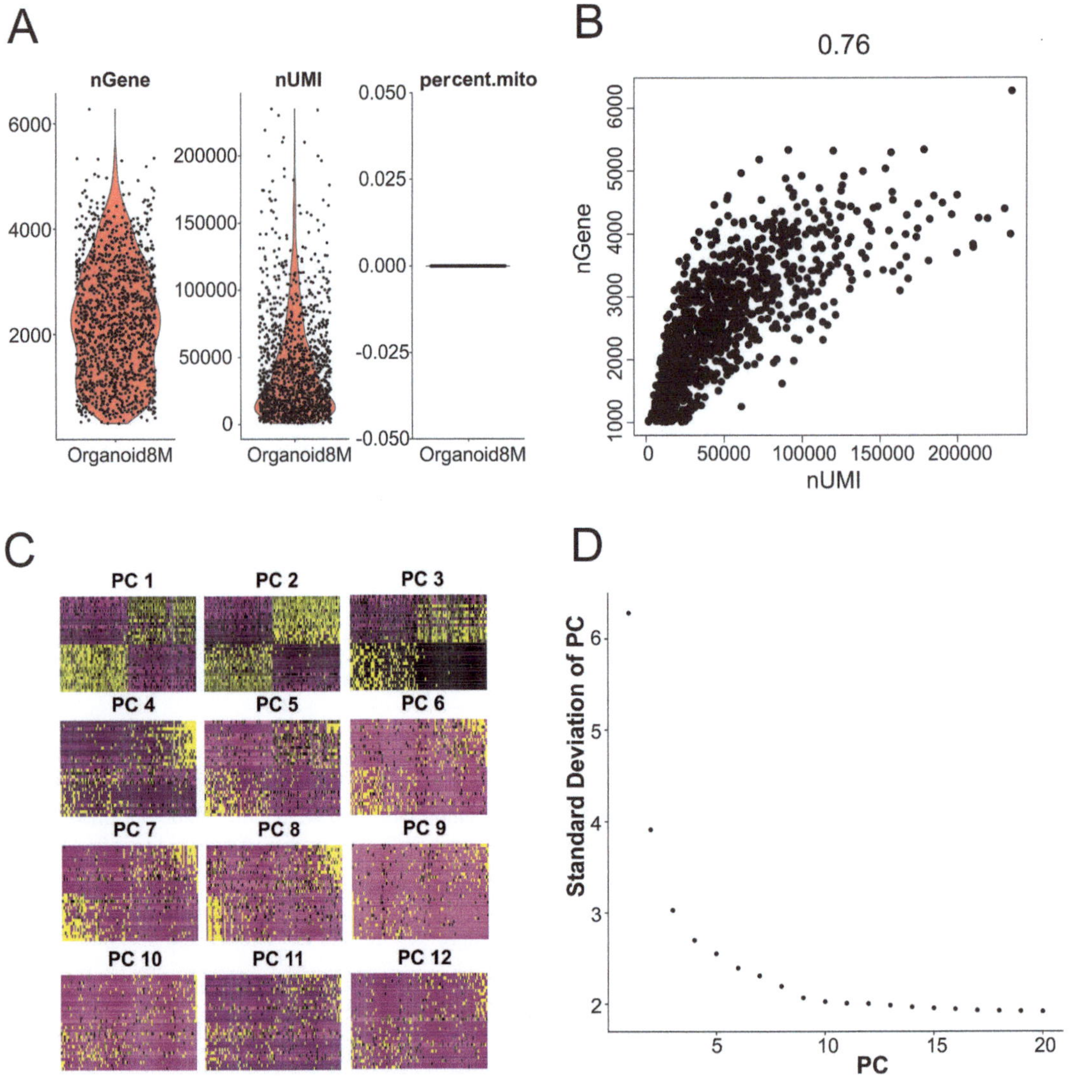

Fig. 6 QC plot of organoid scRNA-seq data. (**a**) Violin plots of the total number of genes, UMIs, and the percentage of mitochondrial genes. (**b**) Scatter plot between nGene and nUMI. (**c**) Heatmap of PCA output. (**d**) PCElbowPlot provides the number of minimal PCs for downstream analysis

2. Install R software.

   In this practice, we use Seurat package running on open-source R environment. First, download and install R (>R-3.5) from R-project (http://cran.revolutionanalytics.com/).

3. Install R packages.

   After installing and opening R, install the packages by typing the following commands in the console panel.

```
source("http://bioconductor.org/biocLite.R")
biocLite(c("Seurat","dplyr","topGo","tidyverse","org.Hs.eg.
db","org.Mm.eg.db"))
```

4. Set a path for the sample data, and load R source and libraries.

   Set your data folder where the organoid data are located, and import the organoid data from a tab-delimited text file. Modify the path ("Input_data_PATH") where the input data and R source are located.

```
setwd("/Input_data_PATH/")
source("Utilities_sc.R")
library(Seurat)
library(dplyr)
library(topGO)
library(tidyverse)
library(org.Hs.eg.db)
library(org.Mm.eg.db)
```

5. Create a data object.

   Here, we use additional options to keep all genes expressed in > = 68 cells (~0.5% of the data) and all cells with at least 300 detected genes.

```
exprTable <-
read.table("scRNAseq_readCount_organoid8M.txt",sep="\t",hea
der=T,row.names=1,quote="")
geneCutoff <- dim(exprTable)[2] * 0.05
dataObj <- CreateSeuratObject(raw.data = exprTable,
min.cells = geneCutoff, min.genes = 300, project =
"Organoid")
```

3.3.2 Quality Check

Data QC can be performed for each sample to ensure high cell viability, high cDNA library complexity, and sufficient sequence coverage. This QC step is critical to assess the quality of the transcriptome for each cell with incomplete cell lysis or failures during library preparation. Here, we can easily assess three attributes in the data: the total number of transcripts detected, the total number of genes found to be expressed, and the fraction of expression contributed by mitochondrial genes. Cells with aberrant behavior are typically removed from further analysis, where we should consider the heterogeneous population of cells and other features can vary substantially [21].

1. Select mitochondrial genes, and calculate the percentage of their overall expression against overall gene expression label.

```
mito.genes <- grep(pattern = "^MT-",
rownames(dataObj@data), value = TRUE)
percent.mito <-
Matrix::colSums(dataObj@raw.data[mito.genes,
```

```
])/Matrix::colSums(dataObj@raw.data)
dataObj <- AddMetaData(object = dataObj, metadata =
percent.mito, col.name = "percent.mito")
```

2. Generate QC plot for the number of genes (nGene), the number of UMIs (nUMI), and the percentage of mitochondrial genes (Fig. 7a) (*see* **Note 12**). In the data processing, nUMI indicates the number of read counts. GenePlot is used to visualize nGene and nUMI relationship (Fig. 7b).

```
VlnPlot(object = dataObj, features.plot = c("nGene",
"nUMI", "percent.mito"), nCol = 3)
GenePlot(object = dataObj, gene1 = "nUMI", gene2 = "nGene")
```

3. Filter out cells with fewer than 500 genes detected and percent.mito higher than 0.1 (10%). We filtered cells by using the "nGene" attribute in "subset.names" in the command line because our data do not include mitochondrial genes, but you can also filter cells by using "percent.mito."

```
dataObj <- FilterCells(object = dataObj, subset.names =
"nGene", low.thresholds = 1000, high.thresholds = Inf)
```

3.3.3 Normalization

After removing unwanted cells from the dataset, the next step is to normalize the data. By default, we employ a global-scaling normalization method "LogNormalize" that normalizes the gene expression measurements for each cell by the total expression, multiplies this by a scale factor (10,000 by default), and log-transforms the result.

1. Normalization using "LogNormalize" option.

```
dataObj <- NormalizeData(object= dataObj,
normalization.method = "LogNormalize", scale.factor =
10000)
```

2. Identify variable genes across cells for downstream analysis. The parameter settings may vary based on the data type, heterogeneity in the sample, and normalization (*see* **Note 13**). Here, we use the defaults.

```
dataObj <- FindVariableGenes(object= dataObj)
```

3. Z-scoring the data using scale factor 10,000 for multidimensional reduction. You can use "vars.to.regress" option to remove uninteresting sources of variation such as "nUMI" and "percent.mito."

```
dataObj <- ScaleData(dataObj)
```

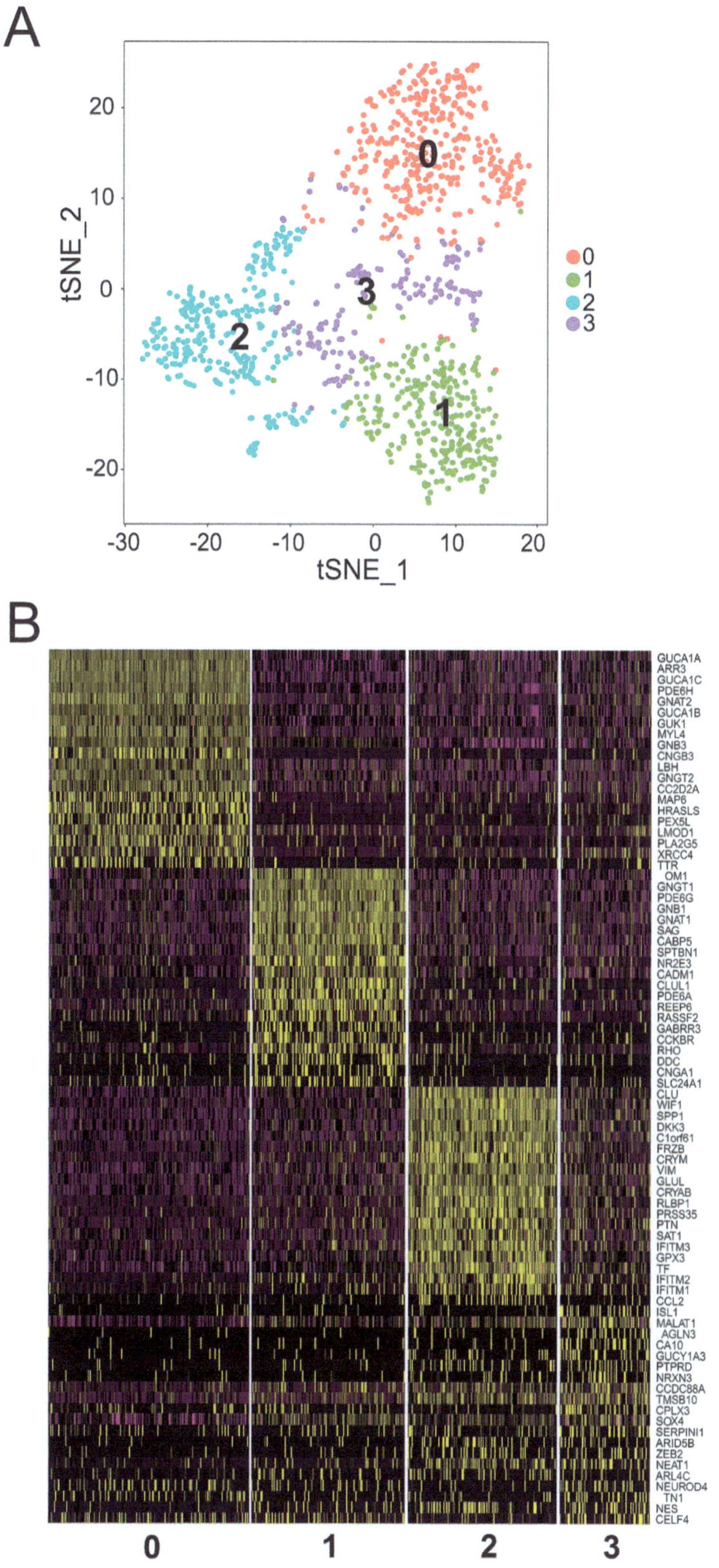

Fig. 7 t-SNE plot and heatmap of organoid scRNA-seq data. (**a**) t-SNE plot of the organoid data using Seurat package. Four subclusters are identified. (**b**) Heatmap of the top 20 genes for each cluster compared with other clusters

3.3.4 Dimensionality Reduction and Visualization

In order to reduce the high-dimensional gene expression matrix to its most important features, Seurat uses principal components analysis (PCA) to change the dimensionality of the dataset from (cells × genes) to (cells × M) where M is a user-selectable number of principal components. After running PCA, for visualizing data in 2-D space, Seurat passes the PCA-reduced data into t-SNE, a nonlinear dimensionality reduction method. The subpopulations of corneal cells are visualized in t-SNE plot.

1. Select highly variable genes (HVGs).

   HVGs are calculated by the variance and mean for each gene in the dataset, and genes are sorted by their variance/mean ratio. Here we use 1000 HVGs for PCA.

```
hvgs <- head(rownames(dataObj@hvg.info), "1000")
```

2. Run PCA.

   The scaled data are used as input for running PCA. HVGS can be defined using "genes.use."

```
dataObj <- RunPCA(dataObj, genes.use=hvgs)
```

3. Visualize PCA output.

```
PCAPlot(object = dataObj, dim.1 = 1, dim.2 = 2)
```

4. Generate the heatmap of PCA output.

   It allows to decide how many PCs are considered for further downstream analyses. Both cells and genes are ordered according to their PCA scores (Fig. 7c).

```
PCHeatmap(object = dataObj, pc.use = 1:12, cells.use = 100,
do.balanced = TRUE, label.columns = FALSE)
```

5. Determine significant PCs.

   Seurat provides JackStraw procedure with a random resampling test to determine statistically significant principal components. In addition, "PCElbowPlot" provides a plot of the standard deviations of the principle components. You can choose a cutoff where there is a clear elbow in the graph (*see* **Note 14**). Here, we use 10 PCs (Fig. 7d).

```
dataObj <- JackStraw(object = dataObj)
JackStrawPlot(object = dataObj, PCs = 1:18)
PCElbowPlot(object = dataObj)
```

6. Run Clustering.

   Clustering based on graph-based clustering approach is performed with the determined number of PCs. You can choose more clusters by increasing "resolution" values.

```
dataObj <- FindClusters(object = dataObj, resolution =
0.6, reduction.type = "pca", dims.use = 1:10)
```

7. Run t-SNE.

   t-SNE plot is used to visualize and explore scRNA-seq data after running "RunTNE." Here, we use 10 PCs which is the same as the input to the clustering analysis (Fig. 8).

```
dataObj <- RunTSNE(object = dataObj, dims.use = 1:10,
do.fast = TRUE)
TSNEPlot(object = dataObj, pt.size = 2, do.label = T,
label.size = 10)
```

3.3.5 Identifying Marker Genes and Assign Cell Type

We can find cluster-specific markers that define clusters via differential expression using "FindMarkers" function which can identify significantly expressed genes in each cluster relative to all other clusters.

1. Find markers for each cluster.

```
dataObj.markers <- FindAllMarkers(object = dataObj,
only.pos = TRUE, min.pct = 0.25, thresh.use = 0.25)
```

2. Select top 20 markers per cluster.

```
outTop20 <- dataObj.markers %>% group_by(cluster) %>%
top_n(20, avg_logFC)
```

3. Draw a heatmap (*see* **Note 15**).

```
DoHeatmap(object = dataObj, genes.use = outTop20$gene,
slim.col.label = TRUE, remove.key = TRUE, group.cex=10,
cex.row=8)
```

4. Test retinal-specific marker expression.

   In the practice, we examined the expression of each known type of retinal cell markers and found cone, rod, and Müller glial (MG) cell clusters (Fig. 8a). ARR3, PDE6H, and GUCA1C are known as cone cell marker genes. SAG, GNAT1, and NR2E3 are rod cells. NFIB, DKK3, and RLBP1 are MG cells. Violin plots for marker genes also can be generated by cluster (Fig. 8b) (*see* **Note 16**).

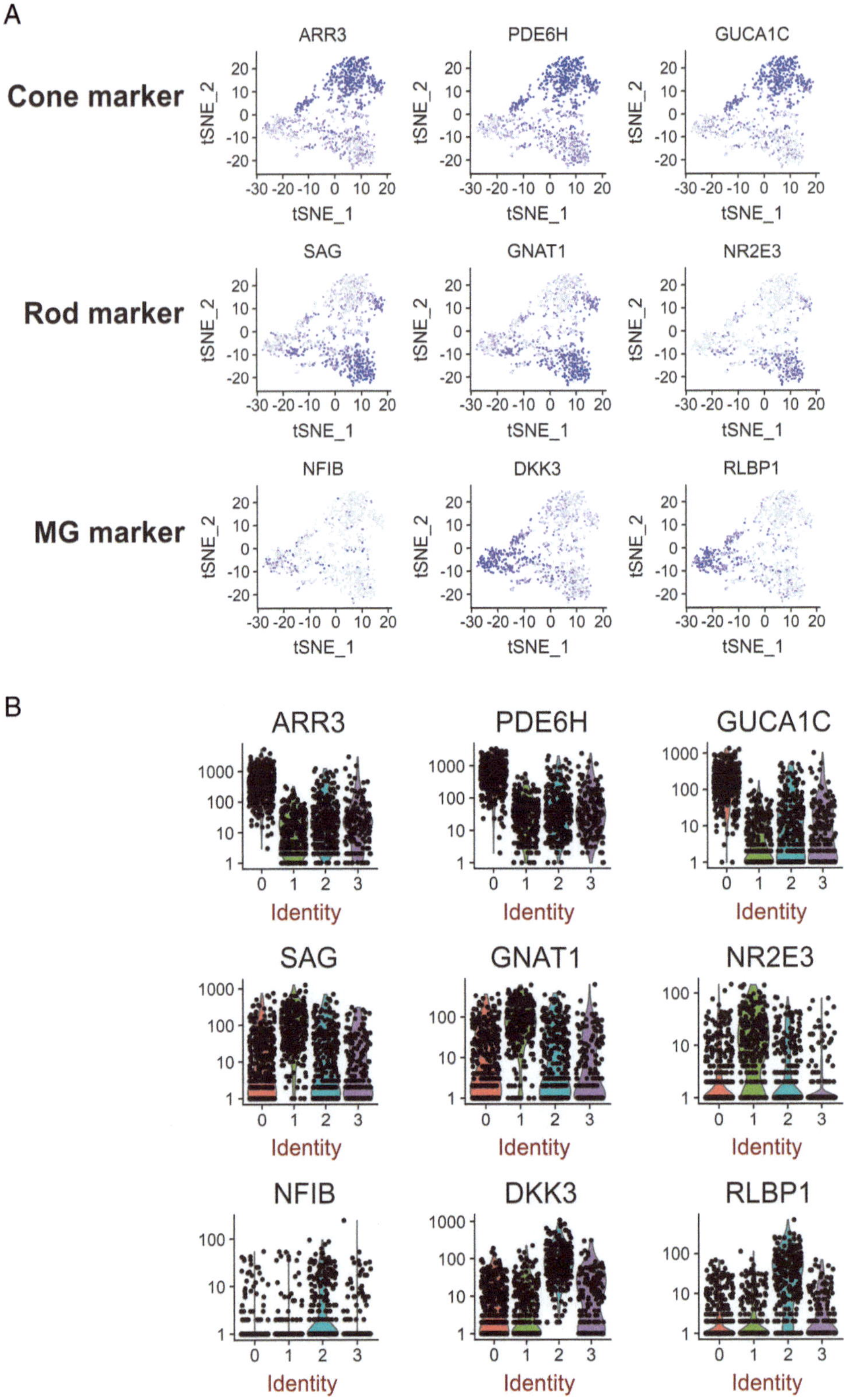

Fig. 8 Expression of retinal-specific markers in each subcluster. (**a**) Feature plots of cone, rod, and MG marker genes. (**b**) Violin plots of cone, rod, and MG marker genes. ARR3, PDE6H, and GUCA1C are used as cone marker; SAG, GNAT1, and NR2E3 as rod; NFIB, DKK3, and RLBP1 as MG. MG, Müller glial cells

```
cellMarker <- c("ARR3","PDE6H",
"GUCA1C","SAG","GNAT1","NR2E3","NFIB","DKK3","RLBP1")
FeaturePlot(object = dataObj, features.plot = geneM1,
min.cutoff="q9", cols.use= c("lightgrey", "blue"),
pt.size=0.5)
VlnPlot(object = dataObj, features.plot = geneM1, use.raw =
TRUE, y.log = TRUE)
```

5. Identify cell type.

   A common step in the analysis of scRNA-seq data is to classify cells into putative subgroups. By identifying these subgroups of cells, the degree of heterogeneity within an interest population can be assessed. As we examined markers in the previous step, the retinal cell types of clusters are assigned and annotate on the t-SNE plot (Fig. 9a).

```
currentClIds <- c(0, 1, 2, 3)
newClIds <- c("Cone", "Rod", "MG", "Mixed")
dataObj@ident <- plyr::mapvalues(x = dataObj@ident, from =
currentClIds, to = newClIds)
TSNEPlot(object = dataObj, do.label = TRUE, pt.size = 2,
label.size=7)
```

3.3.6 Gene Ontology (GO) Analysis

Enriched gene sets of the biological process categories of Gene Ontology can be performed using significant genes among cell clusters. The significant enriched biological processes are identified by p-value less than 0.05 using topGO package in Bioconductor. Enrichment results were plotted (Fig. 9b).

1. Select a cluster and genes for GO analysis.

   Here, we choose cluster-0, assigned as cone cells, and identify top 50 genes of the cluster (*see* **Note 17**).

```
nCluster <- 0
DEGTable <- outTop50[outTop50$p_val<0.01
&outTop50$cluster==nCluster,]
DEGs <- DEGTable$gene
```

2. Run GO analysis (*see* **Note 18**).

```
tGOterms <- topGOterms(fg.genes = DEGs, bg.genes=
rownames(dataObj@data), organism = "Human")
outGOterms <- tGOterms$res.table
```

3. Generate a plot for significant GO terms (*see* **Note 19**).

```
outGO <- outGOterms[,c(2,6)]
GO_term <- outGO$Term
```

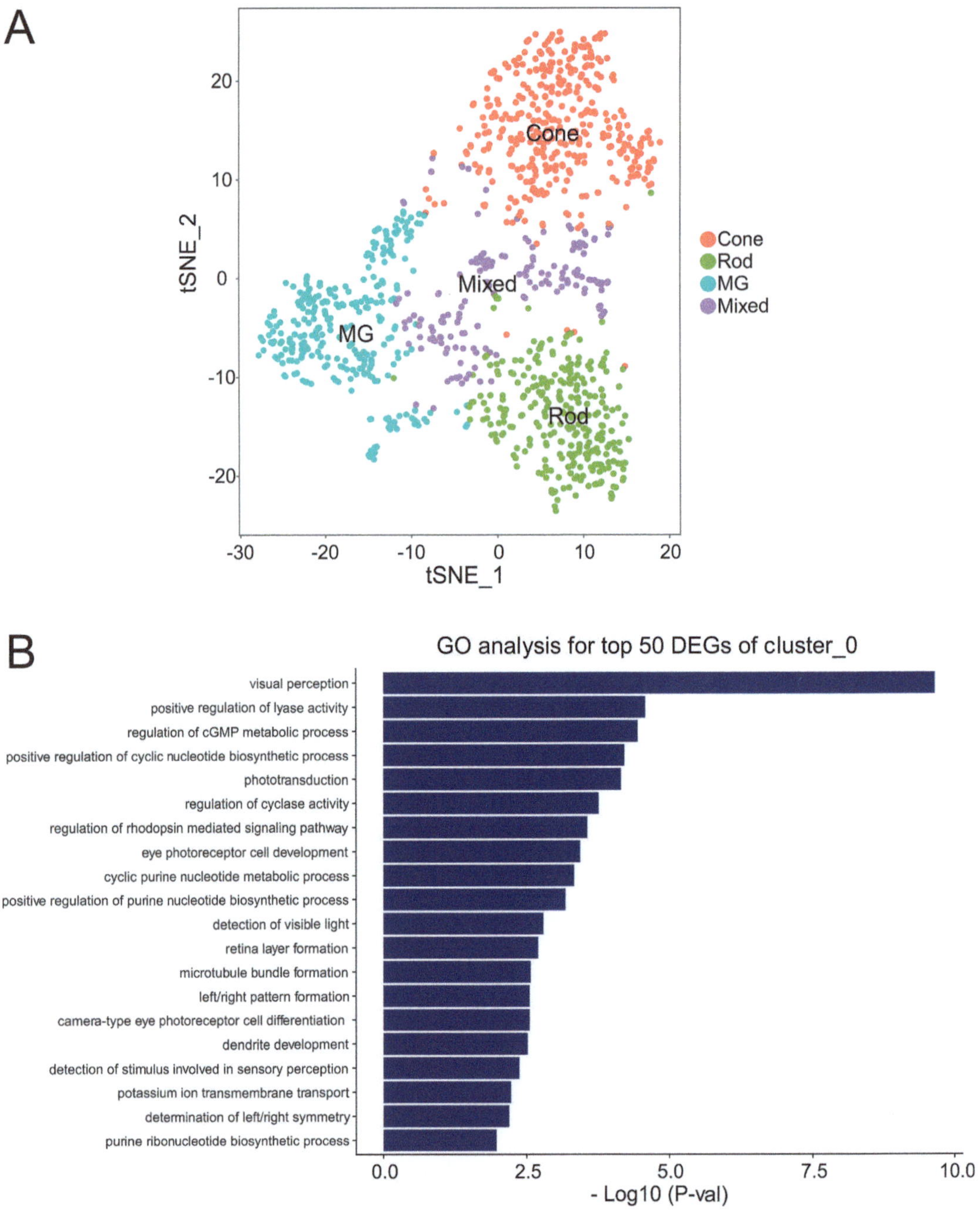

Fig. 9 GO analysis of assigned cone cells. (**a**) Retinal cell types are assigned and visualized to a t-SNE plot based on cell-specific marker expression. (**b**) Enriched biological processes are identified with the top 50 genes in the cone cluster

```
P_val <- -log10(as.numeric(outGO$pval))
bplot <- outGO %>% mutate(GO_term = fct_reorder(GO_term,
P_val)) %>% ggplot(aes(x=GO_term, y=P_val))
```

```
bplot + geom_bar(stat="identity", fill="dark blue") +
coord_flip()+ labs(y="- Log10 (P-val)", x=NULL)+
ggtitle("GO analysis for top 50 DEGs of cluster_0")
```

4 Notes

1. Do not collect more than two eyes at once. Extended incubation on ice/D-PBS will cause the neural retina to shrink.
2. It is important to detach the outer layers including the RPE from the retinal cup very gently. Residual RPE is acceptable if removing it would nick the retinal cup. The eye cup must be intact without tears and nicks.
3. The activation of retinal solution should be timed to completion such that both retinas are freshly dissected for dissociation. In order to achieve this, dissect the first retinal cup and prepare and incubate the papain dissociation solution, and during the incubation time required for papain activation, dissect the second retina.
4. This protocol makes 5 mL of papain dissociation solution which is ideal for dissociation of only two dissected retinal cups. In order to scale up, different reactions can be prepared for each pair of retinal cups, and dissociation can be performed in parallel.
5. The incubation time to digest the retina can vary depending on the age of the mouse and thickness of the retina. Pay close attention to the structure of the eye cup during the swirling process. A good digestion would slightly flatten the eye cup and the result in a wispy retina. The retina should not disintegrate or break while swirling.
6. To test for cell lysis, take 10 μL of solution, mix with 10 μL trypan blue, and visualize under a hemocytometer. Confirm the presence of counterstained nuclei and absence of unstained live cells. If several unstained cells are visible, homogenize the remaining cells 15× again.
7. Mix the cell suspension gently by pipetting several times with a wide-bore pipette to have a homogenous cell suspension.
8. For more information on chip preparation and dispensing procedure, refer to the standard ICELL8 MSND user manual.
9. Equilibrate the DNA beads to room temperature prior to use. Use fresh 80% vol/vol ethanol for washing the beads.
10. This is estimated based on the depth of downstream sequencing required for the specific biological question.

11. When performing data scRNA-seq data analysis, you may try to change parameter settings to dig biological meanings in the data because they can vary depending on the data type, sample heterogeneity, and normalization strategy.
12. At Subheading 3.3.2, **step 2**, you may see the density distribution of "nGene" to choose a cutoff for cell filtering using the following script:

```
hist(dataObj@meta.data$nGene, xlab="nGene").
```

13. At Subheading 3.3.3, **step 2**, you may use different parameter using "x.low.cutoff," "x.high.cutoff," and "y.cutoff" to get other HVGs depending on the relationship between variability and average expression.
14. At Subheading 3.3.4, **step 5**, "JackStrawPlot" provides also significant PCs by random permutation of the subset of data when considering the number of PCs for multidimensional reduction. You may determine the PC number more easily using "PCElbowPlot."
15. At Subheading 3.3.5, **step 3**, you may change the number for getting markers in each cluster. In this practice, we use 20 for heatmap and 50 for GO analysis.
16. At Subheading 3.3.5, **step 4**, based on heatmap (Fig. 7b), you may examine the expression of the significant genes in each cluster using "FeaturePlot" function. In this practice, we classified cluster 0 to 2 as cone, rod, and MG cells based on known retinal cell markers.
17. At Subheading 3.3.6, **step 1**, you can run GO analysis in the different cluster by changing a number of "nCluster" among 0, 1, 2, and 3. And you may test other gene lists which you identify.
18. At Subheading 3.3.6, **step 2**, depending on the species of input data, human or mouse option is available in "topGOterms."
19. During data process, you can save the data object and easily load it back without rerunning heavy computational steps using the following scripts.

```
saveRDS(dataObj, file = "dataObj.rds")
dataObj <- readRDS("dataObj.rds")
```

References

1. Huang S (2009) Non-genetic heterogeneity of cells in development: more than just noise. Development 136(23):3853–3862. https://doi.org/10.1242/dev.035139
2. Shalek AK, Satija R, Shuga J, Trombetta JJ, Gennert D, Lu D, Chen P, Gertner RS, Gaublomme JT, Yosef N, Schwartz S, Fowler B, Weaver S, Wang J, Wang X, Ding R, Raychowdhury R, Friedman N, Hacohen N, Park H, May AP, Regev A (2014) Single-cell RNA-seq reveals dynamic paracrine control of cellular variation. Nature 510(7505):363–369. https://doi.org/10.1038/nature13437
3. Grun D, Lyubimova A, Kester L, Wiebrands K, Basak O, Sasaki N, Clevers H, van Oudenaarden A (2015) Single-cell messenger RNA sequencing reveals rare intestinal cell types. Nature 525(7568):251–255. https://doi.org/10.1038/nature14966
4. Rizvi AH, Camara PG, Kandror EK, Roberts TJ, Schieren I, Maniatis T, Rabadan R (2017) Single-cell topological RNA-seq analysis reveals insights into cellular differentiation and development. Nat Biotechnol 35 (6):551–560. https://doi.org/10.1038/nbt.3854
5. Treutlein B, Lee QY, Camp JG, Mall M, Koh W, Shariati SA, Sim S, Neff NF, Skotheim JM, Wernig M, Quake SR (2016) Dissecting direct reprogramming from fibroblast to neuron using single-cell RNA-seq. Nature 534 (7607):391–395. https://doi.org/10.1038/nature18323
6. Jaitin DA, Weiner A, Yofe I, Lara-Astiaso D, Keren-Shaul H, David E, Salame TM, Tanay A, van Oudenaarden A, Amit I (2016) Dissecting immune circuits by linking CRISPR-pooled screens with single-cell RNA-Seq. Cell 167 (7):1883–1896. e1815. https://doi.org/10.1016/j.cell.2016.11.039
7. Macosko EZ, Basu A, Satija R, Nemesh J, Shekhar K, Goldman M, Tirosh I, Bialas AR, Kamitaki N, Martersteck EM, Trombetta JJ, Weitz DA, Sanes JR, Shalek AK, Regev A, McCarroll SA (2015) Highly parallel genome-wide expression profiling of individual cells using Nanoliter droplets. Cell 161 (5):1202–1214. https://doi.org/10.1016/j.cell.2015.05.002
8. Masland RH (2012) The neuronal organization of the retina. Neuron 76(2):266–280. https://doi.org/10.1016/j.neuron.2012.10.002
9. Rheaume BA, Jereen A, Bolisetty M, Sajid MS, Yang Y, Renna K, Sun L, Robson P, Trakhtenberg EF (2018) Single cell transcriptome profiling of retinal ganglion cells identifies cellular subtypes. Nat Commun 9(1):2759. https://doi.org/10.1038/s41467-018-05134-3
10. Shekhar K, Lapan SW, Whitney IE, Tran NM, Macosko EZ, Kowalczyk M, Adiconis X, Levin JZ, Nemesh J, Goldman M, McCarroll SA, Cepko CL, Regev A, Sanes JR (2016) Comprehensive classification of retinal bipolar neurons by single-cell transcriptomics. Cell 166 (5):1308–1323. e1330. https://doi.org/10.1016/j.cell.2016.07.054
11. Tang F, Barbacioru C, Wang Y, Nordman E, Lee C, Xu N, Wang X, Bodeau J, Tuch BB, Siddiqui A, Lao K, Surani MA (2009) mRNA-Seq whole-transcriptome analysis of a single cell. Nat Methods 6(5):377–382. https://doi.org/10.1038/nmeth.1315
12. Hedlund E, Deng Q (2018) Single-cell RNA sequencing: technical advancements and biological applications. Mol Asp Med 59:36–46. https://doi.org/10.1016/j.mam.2017.07.003
13. Gao R, Kim C, Sei E, Foukakis T, Crosetto N, Chan LK, Srinivasan M, Zhang H, Meric-Bernstam F, Navin N (2017) Nanogrid single-nucleus RNA sequencing reveals phenotypic diversity in breast cancer. Nat Commun 8 (1):228. https://doi.org/10.1038/s41467-017-00244-w
14. Goldstein LD, Chen YJ, Dunne J, Mir A, Hubschle H, Guillory J, Yuan W, Zhang J, Stinson J, Jaiswal B, Pahuja KB, Mann I, Schaal T, Chan L, Anandakrishnan S, Lin CW, Espinoza P, Husain S, Shapiro H, Swaminathan K, Wei S, Srinivasan M, Seshagiri S, Modrusan Z (2017) Massively parallel nanowell-based single-cell gene expression profiling. BMC Genomics 18(1):519. https://doi.org/10.1186/s12864-017-3893-1
15. Zheng GX, Terry JM, Belgrader P, Ryvkin P, Bent ZW, Wilson R, Ziraldo SB, Wheeler TD, McDermott GP, Zhu J, Gregory MT, Shuga J, Montesclaros L, Underwood JG, Masquelier DA, Nishimura SY, Schnall-Levin M, Wyatt PW, Hindson CM, Bharadwaj R, Wong A, Ness KD, Beppu LW, Deeg HJ, McFarland C, Loeb KR, Valente WJ, Ericson NG, Stevens EA, Radich JP, Mikkelsen TS, Hindson BJ, Bielas JH (2017) Massively parallel digital transcriptional profiling of single cells. Nat

Commun 8:14049. https://doi.org/10.1038/ncomms14049

16. Zappia L, Phipson B, Oshlack A (2018) Exploring the single-cell RNA-seq analysis landscape with the scRNA-tools database. PLoS Comput Biol 14(6):e1006245. https://doi.org/10.1371/journal.pcbi.1006245
17. Stegle O, Teichmann SA, Marioni JC (2015) Computational and analytical challenges in single-cell transcriptomics. Nat Rev Genet 16 (3):133–145. https://doi.org/10.1038/nrg3833
18. Dobin A, Davis CA, Schlesinger F, Drenkow J, Zaleski C, Jha S, Batut P, Chaisson M, Gingeras TR (2013) STAR: ultrafast universal RNA-seq aligner. Bioinformatics 29(1):15–21. https://doi.org/10.1093/bioinformatics/bts635
19. Kim D, Langmead B, Salzberg SL (2015) HISAT: a fast spliced aligner with low memory requirements. Nat Methods 12(4):357–360. https://doi.org/10.1038/nmeth.3317
20. Langmead B, Trapnell C, Pop M, Salzberg SL (2009) Ultrafast and memory-efficient alignment of short DNA sequences to the human genome. Genome Biol 10(3):R25. https://doi.org/10.1186/gb-2009-10-3-r25
21. Ilicic T, Kim JK, Kolodziejczyk AA, Bagger FO, McCarthy DJ, Marioni JC, Teichmann SA (2016) Classification of low quality cells from single-cell RNA-seq data. Genome Biol 17:29. https://doi.org/10.1186/s13059-016-0888-1

Chapter 13

Genetically Directed Sparse Labeling System for Anatomical Studies of Retinal Ganglion Cells

Leila Jamal, Takae Kiyama, and Chai-An Mao

Abstract

The stereotypic dendritic morphology is one of the landmark characteristics for classifying retinal ganglion cell (RGC) subtypes. These unique dendritic morphologies and their corresponding stratification level in the inner plexiform layer are indicators of their physiological function and presynaptic connection with other neurons. Mis-patterned dendritic morphologies underlie many neurological disease conditions. To streamline the morphological analysis of RGCs, here, we describe a simple protocol using Cre-/lox-dependent genetically directed sparse labeling strategy on flat-mounted retinas to inspect dendritic morphology of specific RGC subtypes.

Key words Genetically directed sparse labeling, Alkaline phosphatase staining, Neuronal morphology, Cell-type-specific CreER driver, Cre-dependent reporter line, RGC subtypes

1 Introduction

Retinal ganglion cells (RGCs) are the sole output neurons of the retina that collectively transmit visual information to several regions in the brain. In a mature mammalian retina, more than 40 subtypes of RGCs have been described on the basis of their morphological and functional characteristics [1, 2]. While the developmental mechanisms leading to such cellular and functional diversity remain unknown, it is well-recognized that the dendritic morphologies and terminal dendritic stratification of RGC subtypes in the inner plexiform layer correlate with their physiological functions.

Various types of bulk dye loading techniques have been developed to study neuronal morphology [3], such as silver impregnation staining [4], intracellular dye injections [5–8], and multicolor diolistic labeling [9], all of which offer different resolutions and reproducibility outcomes. In the past 15 years, the genetically directed sparse labeling strategy has emerged as a relatively convenient technique to study neuronal morphologies [10, 11]. When combining two (or more) genetically engineered mouse lines,

Chai-An Mao (ed.), *Retinal Development: Methods and Protocols*, Methods in Molecular Biology, vol. 2092,
https://doi.org/10.1007/978-1-0716-0175-4_13,

including a cell-type- or tissue-specific CreER driver and a Cre-dependent reporter line, it can be used for multiple purposes: (1) tracing lineal relationships of progenitor cells, (2) determining neuronal morphology and central projections in the brain, and (3) studying functions of genes of interest (GOI) in specific cell types [11, 12]. There are several advantages to using the genetically directed sparse labeling technique rather than intracellular dye injection. It is high throughput. In a single experiment, the entire spectrum of RGC subtypes in a defined setting can be revealed [13]. It can be used to inspect rare synapse interaction, which is hard to detect using traditional dye filling/recording [14], to trace the central projection of a single RGC [15], to monitor the dendritic development of specific RGC subtypes as well as extremely large neurons [10], to isolate particular populations of RGC types [16], and to study the function of GOI through examining genotype-phenotype relationships when combined with the conditional knockout allele of a GOI [13].

The quality, precision, and clarity of sparse alkaline phosphatase (AP) staining to detect neuronal morphology depend on multiple factors, which are discussed in the notes in this simple protocol. Although our research targets were mouse RGCs, this protocol has broad applications in other systems as well [12].

2 Materials

2.1 Mouse Lines and Mouse Care

These experiments were performed on $Tbr1^{CreERT2/+}$: $Rosa^{iAP/+}$ or $Tbr1^{CreERT2/+}$: $Pou4f1^{CKOAP/+}$ mice of either sex. The $Rosa^{iAP/+}$, $Pou4f1^{CKOAP/+}$, and $Tbr1^{TauGFP\text{-}IRESCreERT2}$ mouse lines were described previously [11, 13, 17] and are illustrated in Fig. 1 (*see* **Note 1**). Preweaning animals were housed with their mother; weaned animals were housed in groups of no more than five in individually ventilated cages. All animal procedures followed the US Public Health Service Policy on Humane Care and Use of Laboratory Animals and were approved by the Institutional Animal Care and Use Committee at The University of Texas Health Science Center at Houston.

2.2 Reagents

1. Tamoxifen to activate CreERT2 recombinase.
2. Ketamine and xylazine to anesthetize the mice.
3. Either 4% paraformaldehyde (PFA) or 10% formalin to fix the mouse eyes.
4. Phosphate-buffered saline (pH 7.4) during retina isolation.
5. AP buffer (100 mM Tris pH 9.5; 100 mM NaCl; 50 mM $MgCl_2$).

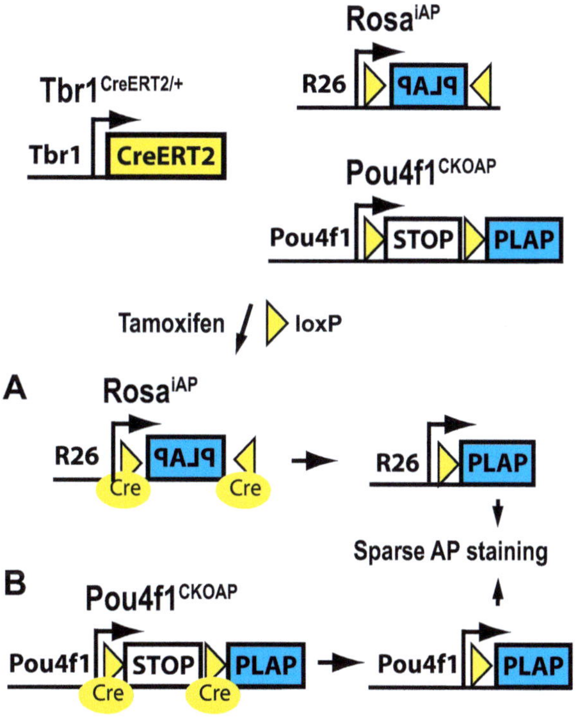

Fig. 1 Schematic illustration of the genetically directed sparse alkaline phosphatase (AP) labeling strategy. (**a**) *Tbr1*$^{CreERT2/+}$:*Rosa*$^{iAP/+}$ and (**b**) *Tbr1*$^{CreERT2/+}$:*Pou4f1*$^{CKOAP/+}$ mouse

6. 4-Nitro blue tetrazolium chloride (NBT; 10 mg/mL) is made by dissolving 0.1 g of NBT powder in 10 mL 30% H_2O/70% dimethylformamide (DMF) mixture. 5-Bromo-4-chloro-3-indolyl phosphate (BCIP; 10 mg/mL) is made by dissolving 0.1 g of BCIP/Na^{2+} in 10 mL ddH_2O (*see* **Note 2**). Small aliquots can be stored at −20 °C indefinitely.
7. AP developing solution: 100 μL of NBT (10 mg/mL) and 50 μL of BCIP (10 mg/mL) are mixed with 10 mL of AP buffer immediately before the AP staining step.
8. Postfix solution: 10% formalin.
9. Clearance solution: benzyl benzoate/benzyl alcohol (2:1, v:v).

2.3 Equipment

1. Dissecting tools: two pairs of fine tweezers (#5) and a pair of Vanna scissors (3 mm blades) for isolating the retina.
2. Small nitrocellulose membrane filters to attach the retina flat mounts.

3. Water bath (65 °C) to inactivate endogenous AP activity.
4. Zeiss Axio Imager 2 microscope equipped with motorized Z drive (Carl Zeiss) for visualizing and imaging fine details of the AP-stained retinal neurons.

3 Methods

3.1 Tissue Fixation and Retinal Extraction

The neurons targeted as described here are RGCs. To obtain precise information for the RGC types studied, the retina must be isolated and flat mounted according to dorsal-ventral and nasal-temporal retinal axis.

1. Tamoxifen (1–5 consecutive daily injections of 100 ug/g body weight) is injected intraperitoneally to label cells.
2. Mice are deeply anesthetized by intraperitoneal injection of ketamine (80 mg/kg) and xylazine (10 mg/kg). Toe pinch response can be used to determine the state of anesthesia. Animals must be unresponsive before continuing. Next, conduct a standard cardiac perfusion with fixative solution (4% PFA). Successful perfusion is determined by tail "twitching" as fixative solution replaces blood within the vessels.
3. After perfusion, poke a tiny hole in the dorsal ora serrata. Carefully remove the eyes and place it in an Eppendorf tube with 4% PFA on ice. Allow the fixative solution to penetrate the tissue for another 5 min.
4. Prepare a small Petri dish with PBS. Carefully pour the eyeballs into the Petri dish. Hold the optic nerve or the tissues that are still attached using a forceps, insert the tip of the Vanna scissors into the hole on the dorsal side, and cut around the ora serrata. Mark the dorsal retina with a tiny cut. After this, carefully pull the lens out using tweezers. To remove the retina that is attached to the retinal pigment epithelium (RPE), carefully make a few cuts from the peripheral rim into the RPE, and gently remove the retina from the RPE by pulling them away from each other. Once the retina is isolated, make 3–4 cuts. Cut a small piece of filter paper, and sink it into PBS. Slowly move the retinas on top, and then use a fine paint brush to flatten the retina onto the filter paper. Move the retina/filter back to the fixative solution for another 5 min at room temperature (*see* **Note 3**).

3.2 Inactivation of Endogenous AP

Endogenous AP activity can be detected in multiple cell types and tissues. Because human placental AP (PLAP) is heat stable, it is critical to first heat inactivate other endogenous AP activity to reduce nonspecific background and enhance the quality of the images.

1. Heat a water bath to 65 °C while preparing retinal flat mounts. Place a beaker with PBS in the 65 °C water bath. After fixation, wash the retina flat mount with PBS. Move the retina flat mount (together with the filter paper) into the beaker with heated PBS for 30 min. This should be sufficient to inactivate any background AP activity yet maintain the PLAP signal in the target population (*see* **Note 4**).

3.3 AP Staining of the AP-Expressing Neuronal Population

1. Remove the retina from heat treatment, and place into a small plastic Petri dish containing AP buffer. Make sure there is enough AP buffer to immerse the retina. Place the Petri dish on a shaker for 5 min.
2. Replace AP buffer with AP developing solution. Place the Petri dish on a shaker for overnight incubation (*see* **Note 5**).

3.4 Clearance of the AP-Marked Neuronal Population

1. Once the intensity of AP staining is satisfactory, replace the AP developing solution with PBS. Shake for 5 min.
2. Replace the PBS with 10% formalin postfix solution. Soak the retina in for 5 min.
3. Replace the postfix solution with a series of ethanol washes (50%, 75%, 90%, 95%, and 100%) of 30 min each.
4. Replace the 100% ethanol with clearance solution.
5. Mount the retina flat mount in clearance solution.
6. Image the AP-marked RGCs with a Zeiss Axio Imager 2 microscope (Figs. 2 and 3).

4 Notes

1. It is critical to select the proper combination of CreER driver line and reporter lines. For example, when combining Jam2-^{CreERT2} with Rosa^{iAP} lines, intense AP staining appears in photoreceptors, amacrine cells, and multiple types of RGCs due to Jam2 expression in multiple cell types. However, when combining Jam2^{CreERT2} with Pou4f1^{CKOAP} mouse lines, most of the AP-marked cells appear to be J-RGCs. In addition, RGCs in Rosa^{iAP} alleles have lower AP activity compared to Pou4f1^{CKOAP} alleles, suggesting that the Rosa26 promoter possesses lower transcription levels than the Pou4f1 promoter in RGCs. In addition, AP staining using the Pou4f1^{CKOAP} line tends to produce more nonspecific AP signal in RGC axons (white arrowheads in Fig. 3), whereas Rosa^{iAP} alleles seem to be "cleaner" and have less background (Fig. 2). Other commonly used AP alleles, such as Z/AP [18], have higher expression levels. However, the Z/AP allele cannot be excised/activated by the Tbr1^{CreERT2} used in this example, possibly due to its less

Fig. 2 Representative image showing the AP staining pattern on retinal flat mounts. (**a**) $Tbr1^{CreERT2/+}$:$Rosa^{iAP/+}$ retina, (**b**) $Tbr1^{CreERT2/+}$:$Pou4f1^{CKOAP/+}$ retina. Note: In $Tbr1^{CreERT2/+}$:$Pou4f1^{CKOAP/+}$ retina, nonspecific AP background staining often appears in axons, but not in the somas. Scale bar: 200 μm

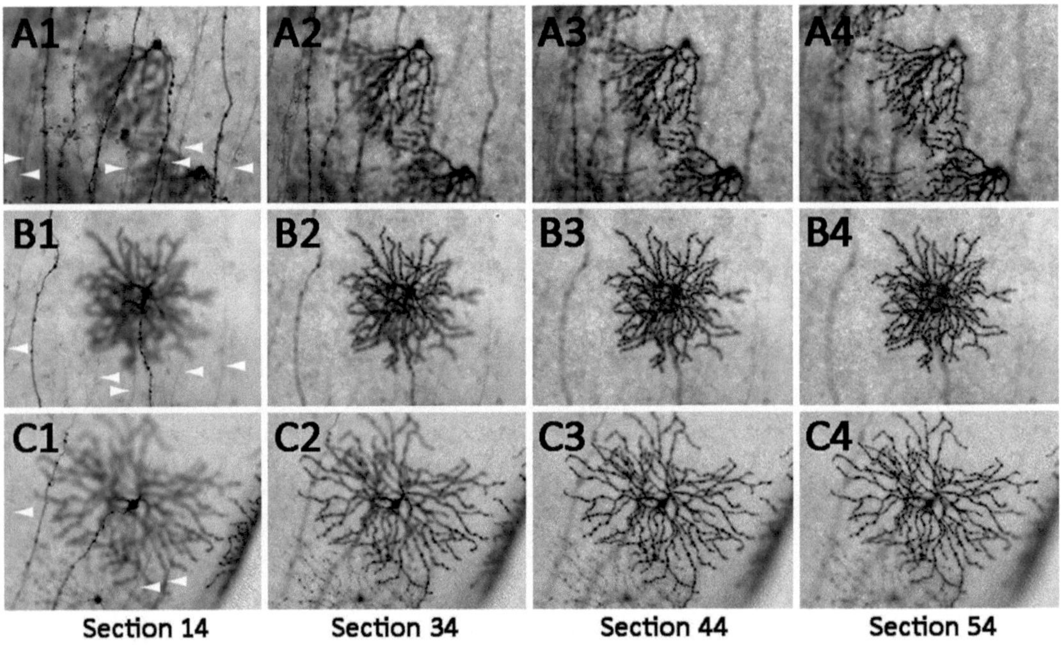

Fig. 3 Representative gallery image showing AP staining on individual Tbr1-expressing RGCs through optic sections. (**a**) An asymmetrical Tbr1-expressing J-RGC, (**b**) a symmetrical Tbr1-expressing OFF RGC with smaller dendritic arbor size, and (**c**) a symmetrical Tbr1-expressing OFF RGC with larger dendritic arbor size. This RGC subtype is reminiscent of the OFF alpha RGCs. Note: The nonspecific AP staining background in $Tbr1^{CreERT2/+}$:$Pou4f1^{CKOAP/+}$ retinas in axons is indicated by white arrowheads

accessible genome structure [19]. The dose and frequency of tamoxifen administration will also affect the number of AP-marked RGCs. For each new experiment, these conditions need to be established first.

2. NBT and BCIP can be purchased from multiple vendors. BCIP is made in different salt formats. The disodium salt form is soluble in water, and the p-toluidine form is soluble in dimethylformamide (DMF).
3. Different reporter lines express different levels of PLAP in different RGC types. Make sure not to over-fix because over-fixation may affect the PLAP enzymatic activity of the target population that has lower levels of PLAP expression.
4. The heat treatment inactivates the endogenous AP present in the tissues. This is the most critical step for obtaining clarity in neuronal morphology because the endogenous AP will generate undesired background noise masking the real signals. Do not use a laboratory hotplate or air-heated oven because the heating in these systems is not even and, therefore, insufficient to eliminate background.
5. The time for developing AP staining depends on the reporter lines and target RGC types. For the $Pou4f1^{CKOAP}$ reporter line, the signal becomes visible in approximately 4–6 h, starting with the soma. The detailed AP-marked dendrites will appear in 12–24 h. For the $Rosa^{iAP}$ line, the signal usually appears in 24 h. It is important to monitor the color development to avoid overdevelopment. However, if the signal is still weak in 24 hours, continue staining for another day.

Acknowledgments

We acknowledge grant support from the National Institutes of Health-National Eye Institute (EY024376 to C.-A.M.), National Eye Institute Vision Core Grant (P30EY028102 to UTHealth), and the Hermann Eye Fund (UTHealth). We acknowledge Dr. Tudor Badea (National Eye Institute) for $Rosa^{iAP}$ and $Pou4f1^{CKOAP}$ mouse lines, Ping Pan (UTHealth) for technical assistance, and Dr. Kimberly Mankiewicz (UTHealth) for manuscript proofreading and editing.

References:

1. Rheaume BA, Jereen A, Bolisetty M, Sajid MS, Yang Y, Renna K, Sun L, Robson P, Trakhtenberg EF (2018) Single cell transcriptome profiling of retinal ganglion cells identifies cellular subtypes. Nat Commun 9(1):2759. https://doi.org/10.1038/s41467-018-05134-3
2. Baden T, Berens P, Franke K, Roman Roson M, Bethge M, Euler T (2016) The functional diversity of retinal ganglion cells in the mouse. Nature 529(7586):345–350. https://doi.org/10.1038/nature16468
3. Parekh R, Ascoli GA (2013) Neuronal morphology goes digital: a research hub for cellular and system neuroscience. Neuron 77 (6):1017–1038. https://doi.org/10.1016/j.neuron.2013.03.008
4. Linke R, Roth G (1989) Morphology of retinal ganglion cells in lungless salamanders (fam. Plethodontidae): an HRP and Golgi study. J Comp Neurol 289(3):361–374. https://doi.org/10.1002/cne.902890302
5. Kong JH, Fish DR, Rockhill RL, Masland RH (2005) Diversity of ganglion cells in the mouse retina: unsupervised morphological classification and its limits. J Comp Neurol 489 (3):293–310. https://doi.org/10.1002/cne.20631
6. Coombs J, van der List D, Wang GY, Chalupa LM (2006) Morphological properties of mouse retinal ganglion cells. Neuroscience 140(1):123–136. https://doi.org/10.1016/j.neuroscience.2006.02.079
7. Coombs JL, Van Der List D, Chalupa LM (2007) Morphological properties of mouse retinal ganglion cells during postnatal development. J Comp Neurol 503(6):803–814. https://doi.org/10.1002/cne.21429
8. Volgyi B, Pan F, Paul DL, Wang JT, Huberman AD, Bloomfield SA (2013) Gap junctions are essential for generating the correlated spike activity of neighboring retinal ganglion cells. PLoS One 8(7):e69426. https://doi.org/10.1371/journal.pone.0069426
9. Gan WB, Grutzendler J, Wong WT, Wong RO, Lichtman JW (2000) Multicolor "DiOlistic" labeling of the nervous system using lipophilic dye combinations. Neuron 27(2):219–225
10. Badea TC, Nathans J (2004) Quantitative analysis of neuronal morphologies in the mouse retina visualized by using a genetically directed reporter. J Comp Neurol 480(4):331–351. https://doi.org/10.1002/cne.20304
11. Badea TC, Cahill H, Ecker J, Hattar S, Nathans J (2009) Distinct roles of transcription factors brn3a and brn3b in controlling the development, morphology, and function of retinal ganglion cells. Neuron 61(6):852–864. https://doi.org/10.1016/j.neuron.2009.01.020
12. Rotolo T, Smallwood PM, Williams J, Nathans J (2008) Genetically-directed, cell type-specific sparse labeling for the analysis of neuronal morphology. PLoS One 3(12):e4099. https://doi.org/10.1371/journal.pone.0004099
13. Kiyama T, Long Y, Chen CK, Whitaker CM, Shay A, Wu H, Badea TC, Mohsenin A, Parker-Thornburg J, Klein WH, Mills SL, Massey SC, Mao CA (2019) Essential Roles of Tbr1 in the Formation and Maintenance of the Orientation-Selective J-RGCs and a Group of OFF-Sustained RGCs in Mouse. Cell Rep 27 (3):900–915. e905. https://doi.org/10.1016/j.celrep.2019.03.077
14. Prigge CL, Yeh PT, Liou NF, Lee CC, You SF, Liu LL, McNeill DS, Chew KS, Hattar S, Chen SK, Zhang DQ (2016) M1 ipRGCs influence visual function through retrograde signaling in the retina. J Neurosci 36(27):7184–7197. https://doi.org/10.1523/JNEUROSCI.3500-15.2016
15. Fernandez DC, Chang YT, Hattar S, Chen SK (2016) Architecture of retinal projections to the central circadian pacemaker. Proc Natl Acad Sci U S A 113(21):6047–6052. https://doi.org/10.1073/pnas.1523629113
16. Sajgo S, Ghinia MG, Brooks M, Kretschmer F, Chuang K, Hiriyanna S, Wu Z, Popescu O, Badea TC (2017) Molecular codes for cell type specification in Brn3 retinal ganglion cells. Proc Natl Acad Sci U S A 114(20):E3974–E3983. https://doi.org/10.1073/pnas.1618551114
17. Badea TC, Hua ZL, Smallwood PM, Williams J, Rotolo T, Ye X, Nathans J (2009) New mouse lines for the analysis of neuronal morphology using CreER(T)/loxP-directed sparse labeling. PLoS One 4(11):e7859. https://doi.org/10.1371/journal.pone.0007859
18. Lobe CG, Koop KE, Kreppner W, Lomeli H, Gertsenstein M, Nagy A (1999) Z/AP, a double reporter for cre-mediated recombination. Dev Biol 208(2):281–292. https://doi.org/10.1006/dbio.1999.9209
19. Long MA, Rossi FM (2009) Silencing inhibits Cre-mediated recombination of the Z/AP and Z/EG reporters in adult cells. PLoS One 4(5):e5435. https://doi.org/10.1371/journal.pone.0005435

Chapter 14

Intracellular Dye Microinjection in Retinal Morphological and Communication Studies

Ye Long

Abstract

The retina provides an excellent system to analyze neural architecture and communication mechanisms. In retinal physiological, morphological, and developmental studies, intracellular dye microinjection is an extensively applied technique. Dyes and/or chemical compounds are introduced into retinal neurons via iontophoresis injection to identify cell types, communication, and developmental changes. Here, we describe the detailed methods for ejecting the dyes Lucifer yellow and Neurobiotin into retinal neurons through micropipette and concentrate on avoiding the known pitfalls and troubleshooting.

Key words Intracellular microinjection, Cell identification, Gap junctions, Neural communication, Lucifer yellow, Neurobiotin, Micropipette, Iontophoresis

1 Introduction

In the studies of structures and functions, especially when the system is as complicated as nervous systems, the most basic questions are what the individual cells look like, what is the relationship between structure and functions, and how they connect to each other in the system network. The great aid for answering such questions may arise from the application of intracellular dye injection which can identify and classify individual cells and understand structural foundation for functions. Microinjection's inception can be traced back to the early 1900s, and this technique boomed from the 1970s. By the 1990s, it had blossomed into a popular technique for biological research [1, 2]. Dye microinjection is the use of a glass of micropipette to inject a small amount of compounds from tubes into living cells. Micropipette production and filling, dye solution preparation, and iontophoresis are critical factors for injection successful rate. In this chapter, we introduce general process and operation for retinal neuron injection with Lucifer yellow (LY) and Neurobiotin (NB). Lucifer yellow (LY) is a fluorescent dye which is widely used as a tracer molecule [3]. Here, we use LY to

Chai-An Mao (ed.), *Retinal Development: Methods and Protocols*, Methods in Molecular Biology, vol. 2092,
https://doi.org/10.1007/978-1-0716-0175-4_14,

visualize the tips of microelectrodes and monitor dye transferring, so low concentration is applied (0.5%). N-(2-Aminoethyl)biotinamide hydrochloride or Neurobiotin (NB) is a useful marker for intracellular tracing, particularly for neurons and the identification of gap junction coupling [4]. Over a century, microinjection techniques have got great progress. It could be widely applied to modern biological researches including cell identification following electrophysiological recording, genetic material introduction, and cell communication via gap junctions, neural pathway tracing, and so on [5–9]. The basic methods of cell injection are similar in different applications and purposes [10]. Thus, by these methods and protocols, you can adjust them for custom using.

2 Materials

Prepare all stock solutions with distilled water before experiments (unless stated otherwise).

2.1 Solutions and Reagents

1. Superfusion solution: Ames' medium, pH 7.4. Weight 8.8 g Ames' medium powder and 1.932 g sodium bicarbonate; add in 1 L distilled water in a graduated conical flask or beaker. Stir until dissolved. Gas the solution with carbogen (95% O_2 and 5% CO_2). Adjust pH with 1 N HCl or NaOH (*see* **Note 1**).
2. Intracellular injection solution: 4% Neurobiotin (NB) tracer (*see* **Note 2**) plus 0.5% Lucifer yellow CH, lithium salt (LY) (*see* **Note 3**) in Tris buffer pH 7.4. Dissolve 1 mg LY and 8 mg NB in 200 μL 0.1 M Tris buffer (*see* **Note 4**).
3. Cell visualizing: Acridine orange (AO) and azure B (AB). Acridine orange is a cell-permeable, nucleic acid binding fluorescent dye for visualizing neurons with epifluorescence microscopy. Azure B can make retinal neurons visualized with bright-field microscopy. Make low-concentration AO and/or AB by eye with fresh Ames' medium.
4. Fixative: 4% Paraformaldehyde (PFA) pH 7.4. Make fresh 4% PFA with 20% PFA in 0.1 M phosphate buffer.
5. Cy3-conjugated streptavidin: Streptavidin-Cy3 stock solution is stocked at 2–8 °C and used in 1:100 dilution in 0.1 M PB with 0.3% Triton X-100 and 0.01% sodium azide.

2.2 Equipment

1. Vibration isolation table and Faraday cage, providing a stable, anti-vibration, and low-noise experimental environment.
2. Microscope: a high-quality upright microscope with a specimen stage which is capable to hold a recording chamber is required. The microscope should have a long working distance condenser, a water emersion 40× objective, and a

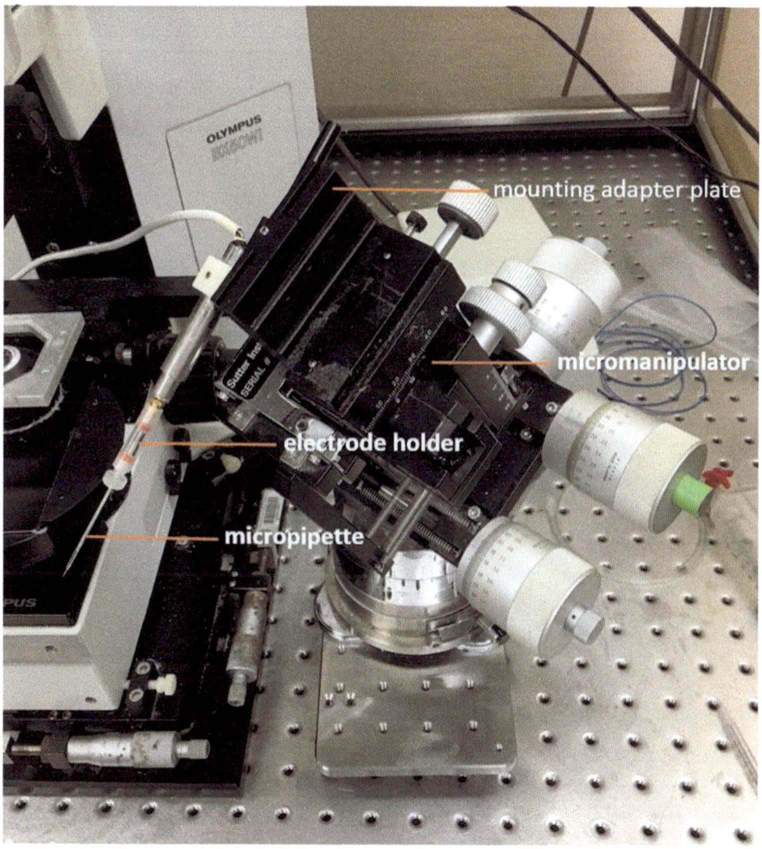

Fig. 1 A headstage fixed on the mounting adapter plate with a micropipette in the electrode holder. Micromanipulator controls precise movements of a headstage

low-magnification (10× or 4×) objective for finding the starting position of a micropipette in the center of view.

3. Headstage: including a micropipette holder with electrode—an Ag/AgCl wire that is coated with a thin layer of silver chloride by dipping in bleach. Headstage is fixed on the mounting adapter plate and holds the micropipette (Fig. 1). When the electrode is submerged into the pipette solution and the ground wire is put in chamber solution, built-in circuit with a current pump (amplifier or microinjector) is completed.
4. Micromanipulator: controlling fine movement under high magnification in the *x*, *y*, and *z* axes.
5. Oscilloscope: monitoring the current pulse during iontophoresis.
6. Illumination sources: for live-cell imaging, a tungsten halogen lamp can provide the field illumination for differential interference contrast (DIC), phase contrast, and so on. The mercury lamp system provides intense illumination over selected

wavelength bands when combined with the appropriate filters for epifluorescence microscopy which is useful for picking out fluorescent-labeled target cells and monitoring dye delivery.

7. Amplifier: most modern amplifiers are equipped with current pump which can eject charged dyes or substances from micropipette into a neuron by generating negative and/or positive current pulses through the circuit.
8. Micropipette puller: glass micropipette is fabricated through heating and pulling processes using a Model P-97 micropipette puller (Sutter Instruments, Novato, CA), which is the lasted generation of filament/brown micropipette puller and has better control of the process.
9. In-line heater: maintains the temperature of the solution in the chamber.

2.3 Accessories

1. Borosilicate glass with filament (OD, 1.2 mm; ID, 0.69 mm; 10 cm length, fire polished) (*see* **Note 5**).
2. MicroFil needle with long and fine tip: filling very close to micropipette tip (*see* **Note 6**).
3. Superfusion chamber.
4. Filter membranes.
5. Superfusion and disposal containers and tubes.

3 Methods

All procedures are conducted at room temperature and in the dim light condition, if possible (*see* **Note** 7).

3.1 Tissue Preparation

1. Dissect an eye of mammals (rabbit, mouse, etc.): hemisect the eyeball, and then harvest the eye cup by removing anterior pole and cleaning up vitreous humor (*see* **Note 8**). Flat the retina by radial cuts or cut the retina in pieces. Tissue can be used with or without sclera.
2. Put the retina in a superfusion chamber which is mounted in an upright microscope with ganglion cell layer facing up. A mesh, harp, or filter paper can be used to prevent retina floating (*see* **Note 9**). Superfuse tissue with oxygenated Ames' medium.
3. Let retina settles down more than 30 minutes before starting experiments (*see* **Note 10**).

3.2 Micropipette Preparation

1. Make a glass micropipette: use micropipette puller with customized program to make an appropriate micropipette (*see* **Note 11**).

2. Tip and back filling of micropipette: fill micropipette tips with 0.5% Lucifer yellow (LY) and 4% Neurobiotin (NB) in Tris buffer. Backfill with 3% LiCl (*see* **Note 12**).

3.3 Equipment Setup

1. Superfusion setup: Ames' medium is oxygenated in an external container which is put on a high stage for gravity feed. Ames' medium is fed into the recording chamber via a connecting tube with consistent perfusion rate (2–4 mL/min) (*see* **Note 13**). An in-line heater is set in this inflow tube close to the recording chamber to control the inflow temperature at 34 °C (*see* **Note 14**). After passing through the tissue, Ames is sucked out from the chamber via an outflow tube to a disposal container. The solution height in the chamber reaches an equilibrium level by in- and outflow (*see* **Note 15**).
2. Microinjection circuit connection: connect microelectrode headstage, amplifier, oscilloscope, isolated pulse stimulator, and adaptive filter [11]. Put ground wire in the chamber to close up the electrical circuit (*see* **Note 16**).

3.4 Cell Visualizing

Prior to injection, several techniques which aid to visualize and identify target cells can be applied.

1. Using white light and interference contrast optics to visualize and identify target cells: with isolated whole-mount retina or tissue slices, retinal neuron somas can be visualized and differentiated by characteristic sizes and shapes.
2. Fluorescent dye—acridine orange (AO): a few drops of the vital dye AO (50 μM) are added into the chamber for prelabeling cells to identify them for subsequent microinjection. Targeted cells can be subsequently injected with Lucifer yellow which is visible with same fluorescent filter set (*see* Note **17**).
3. Nonfluorescent dye—azure B (AB): dripping a few AB in the chamber and stopping superfusion for 1–2 min and then restarting perfusion (*see* **Note 18**).

3.5 Intracellular Microinjection

1. By sticking the electrode in headstage holder into the pipette lumen, the micropipette is connected with the circuit.
2. Find the target cell by DIC or fluorescent microscopy (40×), and move it to the center of view.
3. Find the micropipette in low-magnification objective (4× or 10×).
4. Check the resistance of the pipette: The micro-iontophoresis amplifier used has a built-in feature for measuring the pipette resistance. It evokes brief rectangular test pulses, which can be monitored with a standard oscilloscope.

5. Move micropipette close to the tissue surface, and then switch to the high-magnification water objective.
6. Manipulate the microelectrode tip toward the target cell, and gently break the inner limitation membrane and intercellular tissue till the tip touches target cell membrane (*see* **Note 19**).
7. Break cell membrane with an electrical current by pressing "buzz" button for a few times if needed (*see* **Note 20**). While cell membrane is broken, Lucifer yellow dye can be seen delivered into the cell, and then turn on current pump. Negative and positive currents alternately eject LY (−) and MB (+) into neurons (*see* **Note 21**). Adjust injecting current according to the observation on LY fluorescence intensity. Depending on the sizes and communication of cells, injection duration can last 3–10 min (*see* **Note 22**).

3.6 Fixation and Immunolabeling

After injection, keep retinal tissue in perfusion chamber for more than 20 min for NB transferring to coupled cells via gap junction.

1. Flat the retina on a dark membrane filter, and then put it in 4% PFA for more than 1 h at room temperature.
2. Wash tissue in 0.01 M PBS three times, 20 min for each.
3. Put tissue in Cy3-conjugated streptavidin (1:100) overnight at 4 °C.
4. Flat and mount the retina on slides with mounting medium (VECTASHIELD, Vector Laboratories, Burlingame, CA) for fluorescent imaging.

4 Notes

1. Adjust pH after equilibration with 95% O_2/5% CO_2 for pH might be changed after O_2 and CO_2 dissolve.
2. Neurobiotin (MW 286), an amino derivative of biotin, is combined with LY to visualize neural architecture and identify gap junction coupling [12]. These small molecules are small enough to give a diffusion pattern through gap junction channels for electrical coupling and do not leak across a normal nonjunctional membrane. The common concentration of NB for microinjection is 2–4%. NB-filled cells can be localized by using avidin conjugates. NB has revealed extensive networks of coupled cells in the retina many times larger than seen with LY [13].
3. LY (MW 457) as a fluorescent dye was used in intracellular injection almost a half century ago. Potassium salt can decrease LY solubility; therefore, lithium salt is more suitable for the job. If LY is used alone in injection, higher concentration

(2–4%) can be used. With NB, it is helpful to include a low concentration of LY (≤0.5%) to visualize microelectrode tips and confirm that the injection has been completed. Undissolved substances in the solution can obstruct micropipette, so prior filtration and centrifugation are recommended. Completely dissolve LY/NB in Tris buffer pH 7.4 by using ultrasound cleaner and syringe-driven filter unit (0.22 μm membrane filter) for 2–3 times to avert the insoluble particles which might clog sharp micropipette tips.

4. 0.01 M Phosphate buffer saline (PBS) can be used to replace Tris buffer with pH checking.
5. Several different types of borosilicate glasses are commercially available for the production of micropipettes, including standard (thick-wall) and thin-wall glasses with a variety of outside diameters (OD) and inner diameters (ID). Thin-wall glass can make the micropipette with shorter shank and larger tip diameters which lower the electrical and physical resistance that would aid more dye injected into cells, but the disadvantages of larger tips are increasing the risk of cell damages and the unstable touch between the tip and cell membrane because of the smaller contact area. Micropipette made with thick-wall glass has long shank and much small tip diameter, which give less damage, especially for small neuron penetration; meanwhile, the strong and sharp-tipped thick-wall microelectrodes can easily penetrate the thick and tough intercellular tissue. However, the disadvantage of thick-wall micropipette is that the high resistance and tiny tip might clog the dye injection. The commonly used resistances for thick- and thin-wall micropipettes are 100–150 MΩ and 20–30 MΩ, respectively. The choice of glasses depends on the needs and experience. Glasses with internal filament are popular since the filament increases the capillarity of the micropipette so that solution is drawn into the tip [14].
6. MicroFil needles provide easy way to fill micropipettes, but if used repeatedly and not cleaned thoroughly, they might be obstructed and contaminated. Alternative way is making disposable needle with pipette tips (Fig. 2).
7. Dim light condition is not obliged in microinjection. But if combined with light response recording, dim red light environments and infrared filter in upright microscope are required to lessen photoreceptor bleach during tissue preparation.
8. The residual vitreous humor might block sharp electrode tip and make it hard to reach and break into neurons. Rabbit vitreous usually can be pull out in one piece while the hemisecting cut is more than 1–2 mm far from the corneal limbus. In mice, vitreous is much less and harder to clean up compared

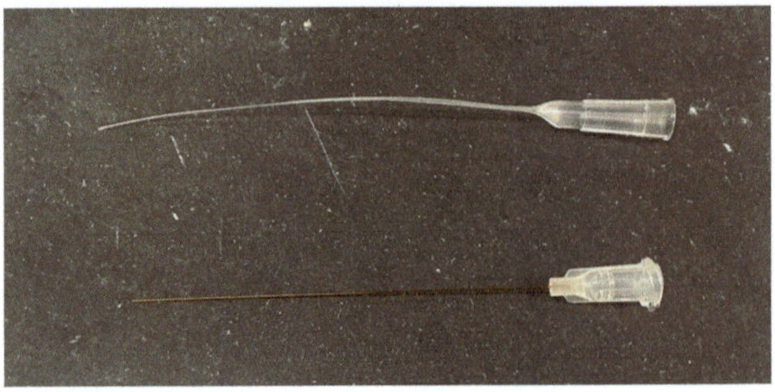

Fig. 2 The comparison of a MicroFil (28GAUGE/67MM) (bottom) and a self-made filling needle by a 200 μL pipette tip (top)

with rabbit. Mouse vitreous is partially adherent to the lens, so do not force to part lens from attachment points. With fine forceps and/or brush, gently pull out the vitreous with lens along the ora serrata.

9. A mesh and/or harp helps settle down the retina in the superfusion chamber (Fig. 3), but mesh and harp can impede microelectrodes to reach cells and then narrow the workable space of the retina. The sclera keeps the retina in a more physiological condition which is beneficial for cell coupling, but it makes retina flatting harder. Flatting the isolated retina on a black filter paper is another way to fix it in the chamber. However, if differential interference contrast (DIC) microscopy will be applied in the experiments, this way is not suitable since filter paper will block the light path.
10. Retinal tissues usually have a little drift after settling down in the chamber, since tissue needs time to fit the new environment and reach a stable condition. During injection, even a bit of shifting and malposition between microelectrode tip and cell can cause the failure of injection.
11. A sharp tip and gradual taper micropipette are a key factor to succeed in microinjection. The pipette puller employs a new strategy for fabrication of micropipettes that enables achieving independent control of their taper and tip diameters. Before designing or adjusting a program with pipette puller, a ramp testing is required to decide the appropriate heat setting while first time using the puller or changing puller filament or glass box. If the heat setting is much higher than the ramp value, there is a risk to burn out the filament. For microinjection, one loop program (heat, pull, velocity, and time) and pressure of air 500 units are commonly applied [15].
12. The potassium salt of LY can form insoluble participate, so electrodes must be backfilled with LiCl when iontophoresing. Some air bubbles in micropipette tips during filling might stop

Fig. 3 Two ways to fix the retina in a chamber. (**a**) A harp and mesh can help to fix and flat the tissue. (**b**) The retina is adhered on a black filter

the dye transportation in injection. Waiting for a few minutes after filling and slightly tapping micropipette may aid small bubbles to come out along capillaries. Sometimes, applying high ejection current a few times can clear small bubbles in the tip as well. If the bubbles are big and close to the tip, discard the micropipette before it causes problems.

13. Perfusion rate can be control by adjusting the height of Ames container and the sizes of inflow tubes.

14. The temperature of superfusion solution is controlled by in-line heater which should be set close to the chamber. However, usually, there is a difference between the actual temperature in the chamber and what is shown in the screen. The difference depends on the sizes of tubes, the distance from

in-line heater to the chamber, and the speed of superfusion. In some cases, the difference can go up to a few degrees Celsius. The set temperature needs to be adjusted to make the actual temperature at 34 °C in the chamber (measuring with thermometer).

15. Perfusion solution height in the chamber should maintain at an immobile and proper level. Too low level can cause the damages of micropipette and tissue by 40× water immersion objective, so don't force the objective to touch the low-level solution; instead, prepare a transfer pipette with Ames' medium and drip a few drops into the chamber to heighten the solution level for water objective immersion. Too high solution level brings out potential risk of overflow which may ruin the whole optical system of microscope. Adjusting the suction tube angle and making an inclined opening of the tube maintain solution surface at a stable level.
16. All electronic equipment should be grounded to reduce noise for monitoring injection current clearly.
17. AO is a nontoxic dye which is easily faded by UV light. Application of AO repeatedly might increase the background light, so the strength and duration of UV light should be reduced as far as possible.
18. Azure B dye intercalates DNA and partitions into the membrane lipid of cells. After staining with AB, cells exhibit variable blue. While identifying neurons, the different characteristics between azure B and AO staining for the same cell types should attract attention.
19. Push the cell a little bit to make a dimple on membrane surface with the micropipette to make sure the tip is in the right place. Limited pressure should be applied on cells to avoid damage to the cell (Fig. 4).

Fig. 4 A micropipette tip filled with Lucifer yellow + Neurobiotin touches a target cell (labeled by acridine orange) membrane with a proper pressure

20. If Lucifer yellow cannot eject into the cell by a "buzz" current, gently tap the back of micromanipulator to help the sharp tip break in. A large short-current impact from current pump might also help break intercellular tissue and cell membrane, but too much force or current can cause dye leakage around and cell damage.
21. The polarity of injection current applied depends on the net charges of the compounds to be ejected. Negative pulse was used to inject negatively charged substance and vice versa. As LY has two negative charges, it can be ejected by passing negative electrical currents. NB can be ejected by positive electrical currents. Continuous current might cause the blockage of micropipette tips. The waveform on oscilloscope can show signs of the block. Sometimes, reversing the polarity of the current or injecting a huge-current impact for a short time can relive the block. Bi-phase of current pulse (+/−) with short interval can reduce the chances of tip clogging. The micropipette should be discarded when the block is irreversible.
22. Current flow is monitored by oscilloscope, and the amount of a substance ejected can be calculated [16]. By observing LY fluorescence brightness, we can roughly estimate the amount, and gently adjust injection current. Usually, a small current with long duration can get better filling than a large current for a short time.

References

1. Stretton AO, Kravitz EA (1968) Neuronal geometry: determination with a technique of intracellular dye injection. Science 162 (3849):132–134
2. Lacal JC, Perona R, Feramisco J (1999). Microinjection. Springer. p. 9. ISBN: 978-3-7643-6019-1. Retrieved 13 July 2013
3. Hanani M (2012) Lucifer yellow—an angel rather than the devil. J Cell Mol Med 16 (1):22–31
4. Huang Q, Zhou D, DiFiglia M (1992) Neurobiotin, a useful neuroanatomical tracer for in vivo anterograde, retrograde and transneuronal tract-tracing and for in vitro labeling of neurons. J Neurosci Methods 41:31–43
5. Mobbs P, Becker D, Williamson R, Bate M, Warner A (1994) Techniques for dye injection and cell labeling. In: Microelectrode techniques. The Plymouth workshop handbook. The Company of Biologists Ltd, Cambridge, UK, pp 361–387
6. Byrne LC, Lin YJ, Lee T, Schaffer DV, Flannery JG (2015) The expression pattern of systemically injected AAV9 in the developing mouse retina is determined by age. Mol Ther 23:290–296. https://doi.org/10.1038/mt.2014.181
7. Hoon M et al (2014) Functional architecture of the retina: development and disease. Prog Retin Eye Res 42:44–84. https://doi.org/10.1016/j.preteyeres.2014.06.003.
8. Zandt BJ, Veruki ML, Hartveit E (2018) Electrotonic signal processing in AII amacrine cells: compartmental models and passive membrane properties for a gap junction-coupled retinal neuron. Brain Struct Funct 223(7):3383–3410
9. Mulderry PK (1996) Intracellular microinjection: a powerful tool in the study of neuronal gene expression and function. Methods 10 (3):292–300
10. Alberto AV, Bonavita AG, Fidalgoneto AA, Berçot F, Alves LA (2017) Single-cell microinjection for cell communication analysis. J Vis Exp 2017(120). https://doi.org/10.3791/50836
11. The axon guide for electrophysiology & biophysics laboratory techniques. Axon Instruments, Foster City, CA

12. Kita A, Armstrong W (1991) A biotin-containing compound N-(2-aminoethyl) biotinamide for intracellular labeling and neuronal tracing studies: comparison with biocytin. J Neurosci Methods 37:141–150
13. Vaney DI (1991) Many diverse types of retinal neurons show tracer coupling when injected with biocytin or Neurobiotin. Neurosci Lett 125(2):187–190
14. Brown KT, Flaming DG (1986) Advanced micropipette techniques for cell physiology. Wiley, New Nork
15. Pipette Cookbook 2018, Adair Oesterle, Sutter Instrument Company, http://www.sutter.com/contact/faqs/pipette_cookbook.pdf
16. Purves RD (1981) Microelectrode methods for intracellular recording and Ionophoresis. Academic Press, New York

Chapter 15

Multielectrode Array Recording of Mouse Retinas Transplanted with Stem Cell-Derived Retinal Sheets

Hung-Ya Tu and Take Matsuyama

Abstract

Retinal multielectrode array (MEA) recording allows us to examine the action potentials of retinal ganglion cells and field potentials of photoreceptors and bipolar cells. In addition to studying the retinal circuitry, it has become one of the standard examination tools for the characterization of stem cell-derived retinal transplantation in degenerated retinas. Besides the detection of responses to simple light stimulation, it is also necessary to consider the spatial correlation of the graft and the electrodes, in order to unbiasedly reveal the locally reconstructed retinal circuitry after transplantation. Here, we introduce our newly developed protocol of MEA recording and analysis that may serve as a standard for evaluating transplanted retinas.

Key words Multielectrode array (MEA), Micro-electroretinogram (mERG), Spike sorting, iPS/ES cell-derived 3D retina, Retinal transplantation, Metabotropic glutamate receptor type 6 (mGluR6), L-AP4, ON bipolar cells, Retinal ganglion cells (RGCs), Bayesian inference

1 Introduction

As one form of ex vivo extracellular electrophysiology, the multi-electrode array (MEA) recording features the ability to detect activities from a large population of cells simultaneously. It has been widely used in studying the retinal ganglion cell (RGC) spiking activities, including light responses of normal retinas [1–4], retinal waves during development [5–7], and spontaneous oscillation in degenerating retinas with photoreceptor loss [8, 9]. In addition to RGCs directly in contact with the electrodes, MEA recording can also detect micro-electroretinogram (mERG), revealing the light-driven ERG a-wave-like hyperpolarization of photoreceptors (mERG a-wave) and ERG b-wave-like depolarization of ON bipolar cells (mERG b-wave) [10]. mERG and spiking activity can be discriminated by signal filters as mERG responses are typically much slower than spiking activity.

In this chapter, we focus on the application of MEA recording of degenerated mouse retinas transplanted with mouse iPS/ES cell-

Chai-An Mao (ed.), *Retinal Development: Methods and Protocols*, Methods in Molecular Biology, vol. 2092,
https://doi.org/10.1007/978-1-0716-0175-4_15,

derived 3D retinal sheets (*see* **Note 1**). The recording of transplanted retinas differs a lot from that of normal, developing (with immature but functional circuitry), and degenerating (with spontaneously active second- and third-order neurons) retinas wherein photoreceptors are evenly distributed or completely lost. The transplants are inserted into the subretinal space and usually only cover a relatively small portion of the host retina, resulting in the locally reconstructed signaling transmission intermingled with spontaneous activities of the degenerated retina. Furthermore, since the 3D retinal sheets transplanted also contain retinal neurons other than photoreceptors, the mERG responses may originate from photoreceptor-bipolar pairs both inside the grafts and between the graft and host. The separation and interpretation of the MEA recording data are therefore critical for objectively evaluating the functional reconstruction of transplanted retinas. The method described here uses simple full-field light stimulation, which demonstrates the light sensitivity (i.e., responses to light onset and offset) as the most essential property of reconstructed retinas by transplantation.

2 Materials

2.1 Reagents

1. Ames' medium.

   To prepare 1 L of Ames' medium (pH 7.4; A1420, Sigma; see the manufacturer website for detailed formulation and instruction), reconstruct the 8.8 g powder mix and 1.9 g sodium bicarbonate by Milli-Q ultrapure water. Other external media (e.g., mammalian Ringer's solution) for acute mammalian retinal explant preparation may also be used alternatively.

2. Opsinamide (AA92593).

   The melanopsin inhibitor is added in the Ames' medium during the entire recording at the working concentration of 10 μM, diluted from the 100 μM stock solution dissolved in dimethyl sulfoxide (DMSO).

3. L-2-Amino-4-phosphonobutyric acid (L-AP4).

   The metabotropic glutamate receptor type 6 (mGluR6) agonist is applied at 10 μM concentration to reversibly block the synaptic transmission from photoreceptors to ON-type bipolar cells. The 100 μM stock solution is prepared in Milli-Q ultrapure water and stored at −30 °C.

4. Isoflurane.

   The inhalational anesthetic is used for deep anesthetization induction and maintenance until cervical dislocation of the animals.

5. 70% Ethanol and Milli-Q ultrapure water.

To clean the perfusion system after the MEA recording.

6. 4% Paraformaldehyde in phosphate buffered solution (PBS).

   The fixative is used for retinal fixation immediately after the MEA recording.

2.2 Equipment for Retinal Preparation

1. Tools for harvesting the retina, removing vitreous, and flat mounting.

   These include straight-pointed iris scissors, 18 Gauge needle, microscissors, fine forceps, transfer pipettes, 35 mm petri dish, absorbent sponge (Medical Quick Absorber), and washer with mesh tightly sealed as the tissue anchor.
2. A dissection microscope equipped with red LED illumination (~690 nm).
3. A light-sealed box for animal transportation.

2.3 MEA Recording and Analysis

2.3.1 Equipment

1. MEA recording system.

   The multielectrode array recording system described as an example in this chapter is MED64 system (Alpha MED Scientific Inc.) with a standard probe (MED-P515A) composed of 8 × 8 electrodes inside the transparent chamber. Each electrode is 50 × 50 μm^2 and spaced 150 μm apart from neighboring ones. There are other systems commercially available with different specification for different animal models, tissues, and purposes.
2. Microscope.

   It is recommended to set up the recording system on an upright microscope equipped with 3 and 4 as listed below. The stimuli can be projected, and images to be taken through the objective lens as the MEA electrodes are usually opaque.
3. White LED stimulation system.

   The stimulation system is synchronized with the recording system for controlling and recording the time of stimulus onset and offset. Detailed examination of retinal properties, such as sensitivities to motion and direction, could be fulfilled by displaying patterns, images, or even movies projected with an OLED projector, for example.
4. IR-DIC imaging system.

   The imaging system allows the experimenter to monitor the retina mounted on the probe during recording and to identify the graft position in correspondence to the electrodes.
5. Peristaltic pump.

   Use a pump with at least two channels to maintain a stable perfusion rate (3–3.5 mL/min) for both in- and outflows to avoid flooding.
6. In-line heater system.

Place the in-line heater as close to the electrode chamber as possible so that the temperature loss may be minimized. The chamber temperature should be monitored independently with a thermal probe. Some commercially available MEA recording systems are built-in with the chamber heater instead.

2.3.2 Software (See Subheading 3.4 for Details)

1. CED Spike 2 (version 7.2).

   Spike sorting (see below for details) can be performed with various package software on the market and/or homemade programs based on MATLAB or Python.

2. R (packages: readxl_1.1.0, zoo_1.8–4, bindrcpp_0.2.2, signal_0.7–6, scales_1.0.0, wesanderson_0.3.6, gridExtra_2.3, forcats_0.3.0, stringr_1.3.1, dplyr_0.7.8, purrr_0.2.5, readr_1.1.1, tidyr_0.8.2, tibble_1.4.2, ggplot2_3.1.0, and tidyverse_1.2.1).
3. RStan (Stan Development Team. 2018. RStan: the R interface to Stan. R package version 2.17.3. http://mc-stan.org).

3 Methods

3.1 Preparation Before Entering the Dark

1. Dark adapt the animal at least 24 h prior to the surgery, and transport it with the light-sealed box before use.
2. Prepare the Ames' medium on the day of recording, and keep it oxygenated (95% O_2 and 5% CO_2) from at least 30 min before and throughout the experiment.
3. Set up the MEA recording system, and pre-clean the perfusion system with MEA chamber installed by running 200–300 mL Milli-Q ultrapure water through. This is also to reduce the risk of flooding or leaking in the dark by making sure the perfusion runs smoothly.
4. Turn on the in-line heater, and monitor the chamber temperature (34–35 °C for the retina).
5. Switch the perfusion from ultrapure water to Ames' medium, and dip the anchor in the MEA chamber in advance of retinal preparation.

3.2 Retinal Preparation (Under Dim Red Light of About 700 nm or Complete Darkness with Night Vision Goggles)

1. Anesthetize the animal with 4% isoflurane in oxygen before sacrificing by cervical dislocation.
2. Enucleate the eyeball by cutting off the connective tissue and optic nerve with straight-pointed iris scissors, and immerse the eyeball in oxygenated Ames' medium.
3. Poke a hole at the limbus with an 18 G needle to cut the cornea out using microscissors, and then remove the lens using fine forceps.

4. Gently isolate the retina from the sclera, carefully avoiding graft detachment from the host retina (*see* **Note 2**).
5. Clean up the isolated retina with fresh Ames' medium.
6. Identify the transplant at the subretinal side by its opaque, spotted appearance (*see* **Note 3**). Remove the vitreous, and trim the retina to a degree that the optic nerve disc and grafted area are retained (*see* **Note 4**).
7. Stop the perfusion to remove the MEA chamber from the system, and take out the anchor. Transfer the piece of the retina to the chamber using a transfer pipette with a cutoff opening.
8. Mount the piece of the retina with the RGC side (epiretinal side) down, and cover the electrodes with the grafted area as much as possible. Remove most of the medium in the chamber, and dry up the surface surrounding the retina with absorbing sponge carefully.
9. Dry up and load the anchor on top of the mounted retina. Carefully install the chamber with the retina back to the MEA system, and restart the perfusion.

3.3 MEA Recording

1. Allow the mounted retina to recover, and adapt to the warmed perfusion medium for at least 20 min before the light stimulation. Monitor the robust RGC spontaneous spiking activity as one of the degenerating retina features.
2. Take photos of the mounted retina using the IR-DIC imaging system for confirming the graft-electrode positioning.
3. Set the low-cut filter to 1 Hz or lower, so that both field potentials as mERG and action potentials mostly from RGCs can be recorded. As shown in the example recording of this chapter (*see* also Figs. 2 and 3), the full-field light stimulation for mERG is of 10 ms duration and 1 s for distinguishing RGCs with different temporal response patterns.
4. Start recordings with preprogrammed light stimuli at different intensities (*see* **Note 5**).
5. Repeat the set of recordings before, during, and after L-AP4 treatment to confirm the dependence of light responses on photoreceptor-bipolar cell synaptic transmission (Figs. 2c,d and 3c). Allow the L-AP4, or any other reagents interested, to be washed in and out for at least 10–15 min, respectively, to ensure the pharmacological effect is fully reversed before the following recordings.
6. After all experiments are done, take photos of the recorded retina with IR and fluorescence (if the transplant is engineered with fluorescence labeling). *See* Fig. 1, for example.

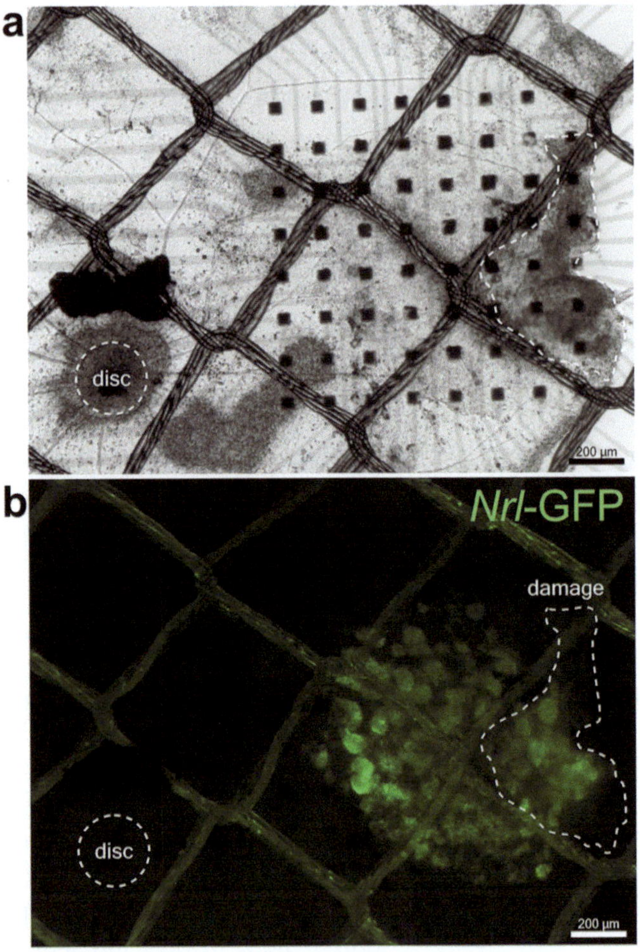

Fig. 1 An example of a transplanted mouse retina mounted on the MEA probe. (**a**) The IR-DIC image of the retina. The retina was stabilized with the mesh-attached anchor. The optic nerve disc was located to the left, and the peripheral retina (darkened region, marked with dotted line) to the right was likely damaged during the transplantation process. (**b**) The *Nrl*-GFP expression of the graft photoreceptors. The graft position can be confirmed and mapped according to the genetically engineered fluorescence observation after the recording. The analyses of mERG waveforms and RGC spikes of this sample are shown in Figs. 2 and 3, respectively

7. Stop the perfusion and take out the retina for fixation. The fixed sample can be preserved and used for immunohistochemical characterization and further analyses of interest.
8. Clean the perfusion system by running 200–300 mL Milli-Q first, followed by 50–10 mL 70% ethanol. Clean the MEA probe and keep it immersed in Milli-Q at 4 °C.

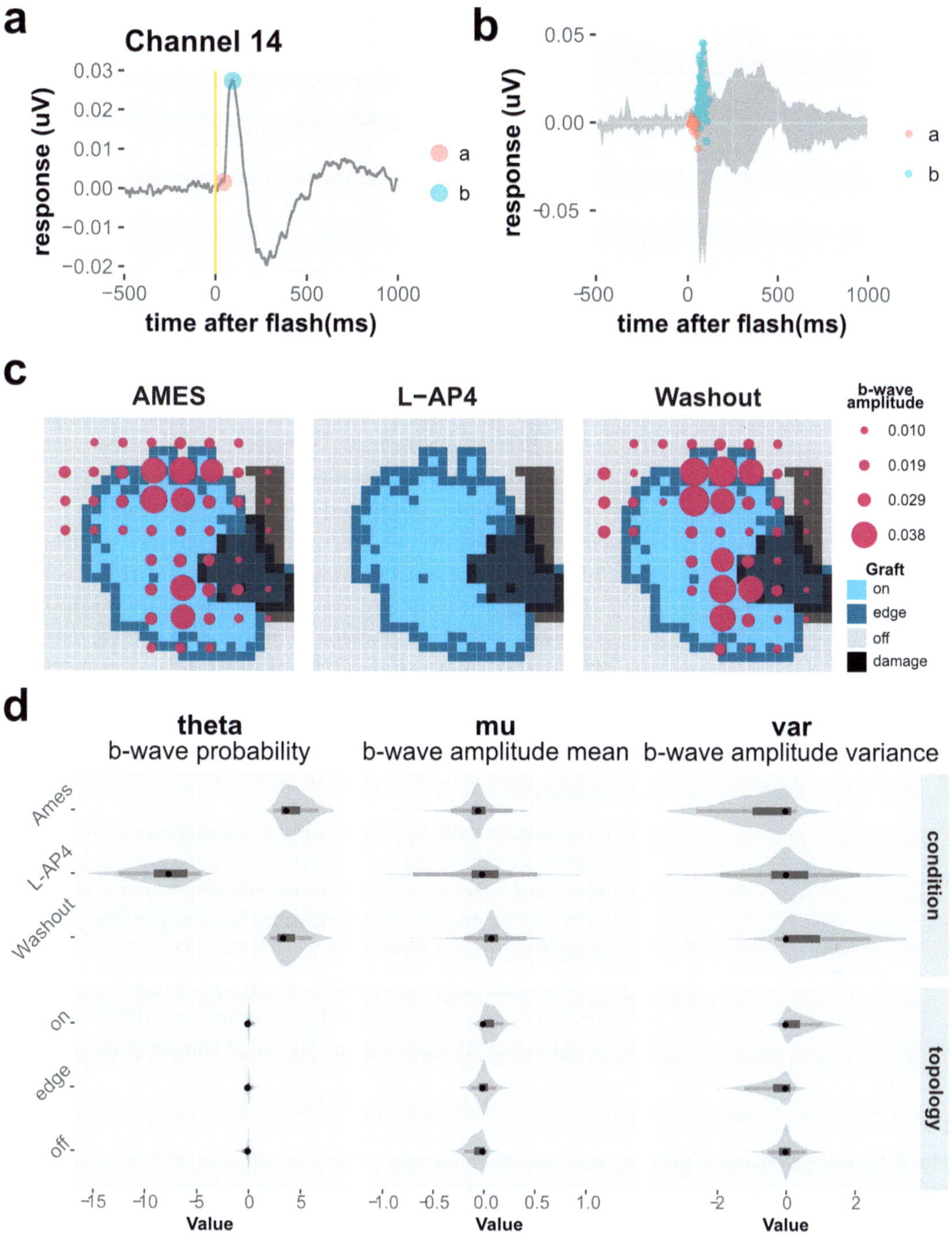

Fig. 2 mERG waveform analysis in correlation to the graft location on the MEA probe. mERG recording of the retina shown in Fig. 1 stimulated with 10 ms full-field light pulse at the intensity of 12.84 log photons/cm^2/s. (**a**) The detection of a-/b-waves in response to the light pulse in a single electrode. (**b**) Overlay of a-/b-wave peaks detected in the entire MEA probe (64 electrodes). (**c**) Distribution of b-wave amplitudes, shown as circles of different sizes, mapped on the probe together with the graft location overlaid and damaged region indicated. The b-waves detected in control (Ames) condition were diminished in the presence of L-AP4 and

3.4 Data Processing and Analysis

1. Graft mapping.
   (a) Map the graft location according to the physical dimensions of the probe. For example, in recordings using an 8×8 MEA probe, graft location is mapped on a 26×26 grid – each square is sized at $50 \times 50\ \mu m^2$, same as the electrode size or half of the interval between neighboring electrodes (*see* **Note 6**).
   (b) Assign each square of the 26×26 grid to indicate whether it is covered ("on") or not ("off") by the graft (*see* **Note 7**).
   (c) Assign each square of the 26×26 grid to indicate whether it is located in "damage" region (*see* Figs. 2b and 3c, for example).
2. mERG a-/b-wave analysis (*see* **Note 8**).
   (a) Set 1–50 Hz band-pass filter to isolate the mERG waveforms from high-frequency RGC spikes and low-frequency noises featured in degenerated retinas.
   (b) Identify all the local maxima and minima within 1 s after 10 ms light stimulation (*see* **Note 9**) by lightly smoothing data and detecting maxima or minima in rolling margins of the time series (Fig. 2a, b; *see* **Note 10**).
   (c) Candidate a-waves are local minima within 55 ms and b-waves are local maxima within 120 ms. Assign the largest peaks within the time window as tentative a- and b-waves.
   (d) Use the recording up until light stimulation onset as the baseline. The offset of baseline from zero and its standard deviation are used later as a reference to identify and correct the magnitude of meaningful changes from chance fluctuations (*see* **Note 11**).
   (e) Plot the heatmap of b-wave amplitude in correlation with the graft location for each recording condition (Ames, L-AP4, washout; see also Fig. 2c, for example); this will

Fig. 2 (continued) recovered following washout. (**d**) The results of Bayesian inference of b-wave amplitude using a Gamma hurdle model. θ represents the probability of generating a b-wave (in inverse logit scale); μ and σ^2 represent the mean and variance of the b-wave amplitude, respectively (in exponential scale). The violin plots show the distribution of credible values estimated by Bayesian inference. The points represent the most likely value, thin light-gray bars represent the 89% interval of the most likely values, and the thick dark-gray bars represent the 50% interval of the most likely values. For simplicity, data from channels in the "damage" region was excluded from this analysis. When processing multiple samples, "damage" may also be added as a separate predictor to the model that estimates the effect of damage. Note the b-wave was not detected in the presence of L-AP4, and therefore, there is great uncertainty in the estimation of mean and variance for this condition

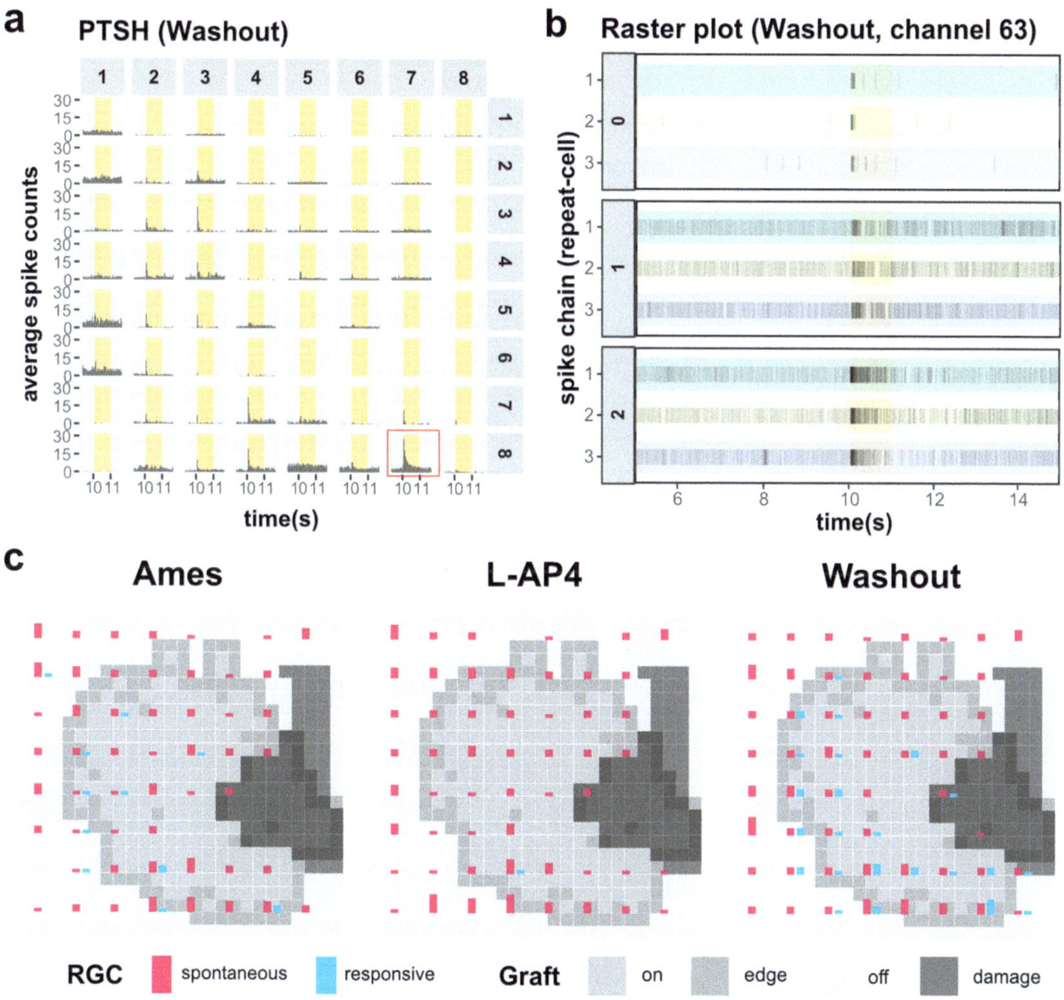

Fig. 3 Graft-driven light responses in host RGCs. Retinal ganglion cell spikes recorded from the retina shown in Fig. 1 with 1 s full-field flash light stimulation at the intensity of 12.84 log photons/cm^2/s. (**a**) Peri-stimulus time histogram (PSTH) of spiking activity at each channel of the MEA probe. The PSTH is shown centered at the 1 s light stimulation as shown by the yellow bar in each channel. (**b**) An example of the raster plot for channel 63 in (**a**), where three cells can be identified with evoked spiking activities correlated to the light stimulus onset. Similar spiking temporal patterns in response to the light stimulation are reproduced in all three repeats of recordings for these cells. (**c**) An overlay plot of the graft location and spiking activity count tally per channel under three different conditions. Gray boxes show "on," "edge," or "off" graft locations. Pink bars indicate the number of cells with activity independent of light stimuli (spontaneous). Blue bars show the number of cells responsive to the light stimulus. The height of the bars is normalized to the largest count in the data set, which is seven in this case. The spontaneous activity is present both in "on" and "off" the graft-covered channels. Responsive cells, on the other hand, seem to be constrained to electrodes covered by the grafted area

provide an overview of the graft-driven bipolar cell activity distribution.

(f) Analyze the mERG b-wave probability, amplitude mean, and amplitude variance among channels, conditions, and samples (*see* **Note 12**). Note the mERG b-waves may originate from bipolar cells of both the transplanted 3D graft and host retina.

3. RGC spiking activity analysis.
   (a) Import the MEA raw data into Spike 2. Apply 200–2800 Hz Butterworth band-pass filter to isolate the RGC spikes, and then separate the spikes from different spiking units (i.e., RGCs) in each channel by spike templates (*see* **Note 13**). The spike sorting results are exported and converted to the xdr format for the following processing with R (*see* **Note 14**), together with the information of onset/offset of recording and stimulation if recorded.
   (b) Make the peri-stimulus time histogram (PSTH) of spike frequency for each channel, as this gives a rough idea of the response types and shows the overall activity at a glance (Fig. 3a).
   (c) Create the raster plot of all sorted spikes, lining up the spike trains of each spiking unit (cell) in each channel (cluster of cells) from multiple repeats of recording (Fig. 3b).
   (d) Summarize the spiking activity of each cell at each channel in each repeat during key time windows, which include before light stimulation (baseline activity), early and late phases of light exposure (response related to light onset), and after light stimulation (response related to light offset). Classify the RGC types based on their spiking activity and temporal pattern of light response (*see* **Note 15** for details) with robustness (by showing agreement with the assignment of response type among repeated recordings).

4 Notes

1. The MEA recording and data analysis described in this chapter have been at least partially reported in our recent studies with allo- and xenotransplantation with different iPSC/ESC retinas transplanted to different animal models [11, 12]. Differentiation of 3D retinal sheets from mouse iPS/ES cells and subretinal transplantation technique are not included in this chapter but can be found with details in our recent reports [13, 14].
2. The graft is located between the retina and the retinal pigment epithelium (RPE), and it is sometimes encapsulated with a semitransparent viscous substance. The graft may be loosely attached to the retina; instead of peeling the retina off, it may

be better to remove RPE together with the retina at first, as the retinal tissue around the graft may be damaged and fused with the RPE due to the transplantation surgery.

3. The graft can sometimes be hardly found under a low-magnification microscope with dim red illumination; however, the whitish spots seen inside the grafted area are rosette structures formed mainly by the graft photoreceptors and can be identified more easily after allowing the freshly isolated retina to rest in clean medium for a short while.
4. The vitreous in the degenerated, aged eye may be slightly dried up and tightly stuck to the epiretinal surface; removing it with fine forceps without damaging the retina may be difficult at first but is critical for having a good contact between the retina and electrodes.
5. It is recommended to repeat each stimulus for multiple times to reproduce the response, as the degenerating retinas are usually hyperactive, and the untimely spontaneous activities may sometimes be misinterpreted as the evoked activities. Also, the spontaneous activities are often found enhanced after repeats of light stimulation.
6. In recordings of transplanted retinas, grafts are rarely distributed over the entire probe – some electrodes are covered by the grafted retina and some are not. It is therefore important to map the location of the graft on the electrode array to distinguish responses from channels with or without graft. MEA probes are typically available in a square grid pattern, and their electrodes are separated by a distance usually larger than the area of individual electrodes. In this example, we show a recording with the 8×8 probe, but the graft location is mapped on a 26×26 grid, considering the distance between the electrodes. Mapping the location of the grafts that are not directly over electrodes is important because these may affect surrounding electrodes.
7. It is useful to define a third "edge" region in addition to "on" and "off" graft. This can be calculated from the "on" mapping by dilating a single unit using a morphological transformation and subtracting this from the "on" mapping.
8. In order to detect the a- and b-waves, each channel is processed separately.
9. mERG can only be analyzed when the light stimulus is short enough, typically $\leqq$ 10 ms, for that responses of photoreceptors, second-order neurons (bipolar cells), and third-order neurons (bipolar cells, ganglion cells, and others) can be distinguished without overlapping. If multiple repeats are recorded, averaging over them can increase the signal-to-noise ratio for peak detection in mERG waveforms.

10. For local minima and maxima detection, some tweaking is required to adjust the degree of smoothing and rolling margin width parameters, depending on the sampling rate and sensitivity of the MEA recording system.
11. It is important to stress that these peaks may not in fact reflect a- and b-waves. We have observed that large negative peaks resembling a-wave are likely strong RGC responses, as these peaks are often sensitive to L-AP4. We have discussed the tentative a-wave candidates to illustrate the potential to analyze them; however, we have rarely observed L-AP4-insensitive a-waves in transplanted retinas, and given that a-wave-like RGC responses are fairly common, the amplitude of the b-wave, which is typically calculated from the value of the a-wave, is calculated from the baseline value instead.
12. We use Bayesian parameter estimation rather than significance testing. Using Bayesian inference, we are able to model complex models and estimate the credible parameter intervals in accordance with the data. For example, the mERG data can be comprehensively modeled with a hurdle model that combines both the probability of generating a b-wave and the amplitude of the b-wave. The absence or presence of b-wave is a binary outcome, and so it is modeled as a logistic regression. The b-wave amplitude, on the other hand, is a continuous value greater than zero, and so it can be modeled with the Gamma distribution. Combining these two, we obtain a Gamma hurdle model:

$$y_n \sim \begin{cases} \text{Bernoulli}(\theta_n) & \text{if } y_n = 0 \\ \text{Gamma}(\mu_n, \sigma_n^2) & \text{if } y_n > 0 \end{cases}$$

where y is the b-wave amplitude, θ represents the probability of generating a b-wave, μ represents the mean b-wave amplitude, and σ^2 represents the variance of the amplitude. In this particular example, the different treatment conditions (Ames, L-AP4, or washout) and the topology of the graft (on, edge, or off) can be used as predictors.

$$\theta_n = \theta_0 + \theta_{\text{cond}} + \theta_{\text{topo}}$$

$$\mu_n = \mu_0 + \mu_{\text{cond}} + \mu_{\text{topo}}$$

$$\sigma_n^2 = \sigma_0^2 + \sigma_{\text{cond}}^2 + \sigma_{\text{topo}}^2$$

In this case, we can see that L-AP4 treatment drastically reduces the chances generating a b-wave, and "on" and "edge" channels show a much more robust response than channels "off" the graft. Note that this is but an example from a single mouse illustrating how data can be analyzed. As the results of transplantation are highly variable, several animals should be analyzed to make inferences, with a parameter accounting for the bias of each sample. With more animals, we see a much more evident trend in "topology," with b-waves observed more often "on" or at the "edge" of the graft. RGC spiking activities can also be modeled similarly. Bayesian parameter inference was implemented using RStan.

13. For spike sorting, interactive templates are created based mainly on the spike amplitude and shape, which may vary depending on electrode impedance, contact between the tissue and MEA probe, the proximity of the cell to electrodes, size of the cell, etc., during recursive scanning of the full recording length. Other factors such as the number and size of templates should be further adjusted according to the sensitivity of the recording and the robustness of the spiking activity.
14. Scripts for processing and analyzing mERG and RGC spiking activity are available at https://github.com/matsutakehoyo/MEA-recording-of-transplanted-mouse-retinas with the sample data shown in this chapter.
15. The baseline activity is obtained by calculating the spike frequency of the 10 s recording prior to light exposure. The duration of light exposure and the window after light stimulation are divided to discriminate different response types. The responses can be categorized to ON (including transient and sustained ON, while delayed ON may be attributed to residual melanopsin-driven RGC intrinsic light response), OFF (including transient OFF and ON suppression), ON-OFF, etc., or the cells with robust spiking activity but not clear light responses are designated as "spontaneous." For example, if the cell exhibits high activity in the time window right after light stimulation (of at least twice as large as baseline activity), this spike train is considered "on" type. If this response persists across repeats, then the cell is assigned that response type. Figure 3c shows an example of such analysis under three different conditions; for simplicity, only the total number of responsive cells is shown ignoring their types.

Acknowledgments

We would like to thank the Laboratory for Retinal Regeneration in RIKEN, Japan, especially for the retinal transplantation team led by Dr. Michiko Mandai. This methodology would not have been developed without the generous support from the team members.

References

1. Elstrott J, Anishchenko A, Greschner M, Sher A, Litke AM, Chichilnisky EJ, Feller MB (2008) Direction selectivity in the retina is established independent of visual experience and cholinergic retinal waves. Neuron 58(4):499–506. https://doi.org/10.1016/j.neuron.2008.03.013
2. Field GD, Gauthier JL, Sher A, Greschner M, Machado TA, Jepson LH, Shlens J, Gunning DE, Mathieson K, Dabrowski W, Paninski L, Litke AM, Chichilnisky EJ (2010) Functional connectivity in the retina at the resolution of photoreceptors. Nature 467(7316):673–677. https://doi.org/10.1038/nature09424
3. Farajian R, Pan F, Akopian A, Volgyi B, Bloomfield SA (2011) Masked excitatory crosstalk between the ON and OFF visual pathways in the mammalian retina. J Physiol 589(Pt 18):4473–4489. https://doi.org/10.1113/jphysiol.2011.213371
4. Tu DC, Zhang D, Demas J, Slutsky EB, Provencio I, Holy TE, Van Gelder RN (2005) Physiologic diversity and development of intrinsically photosensitive retinal ganglion cells. Neuron 48(6):987–999. https://doi.org/10.1016/j.neuron.2005.09.031
5. Firth SI, Wang CT, Feller MB (2005) Retinal waves: mechanisms and function in visual system development. Cell Calcium 37(5):425–432. https://doi.org/10.1016/j.ceca.2005.01.010
6. Wong RO (1999) Retinal waves and visual system development. Annu Rev Neurosci 22:29–47. https://doi.org/10.1146/annurev.neuro.22.1.29
7. Wong RO, Meister M, Shatz CJ (1993) Transient period of correlated bursting activity during development of the mammalian retina. Neuron 11(5):923–938
8. Stasheff SF (2008) Emergence of sustained spontaneous hyperactivity and temporary preservation of OFF responses in ganglion cells of the retinal degeneration (rd1) mouse. J Neurophysiol 99(3):1408–1421. https://doi.org/10.1152/jn.00144.2007
9. Barrett JM, Degenaar P, Sernagor E (2015) Blockade of pathological retinal ganglion cell hyperactivity improves optogenetically evoked light responses in rd1 mice. Front Cell Neurosci 9:330. https://doi.org/10.3389/fncel.2015.00330
10. Fujii M, Sunagawa GA, Kondo M, Takahashi M, Mandai M (2016) Evaluation of micro electroretinograms recorded with multiple electrode array to assess focal retinal function. Sci Rep 6:30719. https://doi.org/10.1038/srep30719
11. Iraha S, Tu HY, Yamasaki S, Kagawa T, Goto M, Takahashi R, Watanabe T, Sugita S, Yonemura S, Sunagawa GA, Matsuyama T, Fujii M, Kuwahara A, Kishino A, Koide N, Eiraku M, Tanihara H, Takahashi M, Mandai M (2018) Establishment of immunodeficient retinal degeneration model mice and functional maturation of human ESC-derived retinal sheets after transplantation. Stem Cell Reports 10(3):1059–1074. https://doi.org/10.1016/j.stemcr.2018.01.032
12. Tu HY, Watanabe T, Shirai H, Yamasaki S, Kinoshita M, Matsushita K, Hashiguchi T, Onoe H, Matsuyama T, Kuwahara A, Kishino A, Kimura T, Eiraku M, Suzuma K, Kitaoka T, Takahashi M, Mandai M (2018) Medium- to long-term survival and functional examination of human iPSC-derived retinas in rat and primate models of retinal degeneration. EBioMedicine. https://doi.org/10.1016/j.ebiom.2018.11.028
13. Assawachananont J, Mandai M, Okamoto S, Yamada C, Eiraku M, Yonemura S, Sasai Y, Takahashi M (2014) Transplantation of embryonic and induced pluripotent stem cell-derived 3D retinal sheets into retinal degenerative mice. Stem Cell Reports 2(5):662–674. https://doi.org/10.1016/j.stemcr.2014.03.011
14. Mandai M, Fujii M, Hashiguchi T, Sunagawa GA, Ito SI, Sun J, Kaneko J, Sho J, Yamada C, Takahashi M (2017) iPSC-derived retina transplants improve vision in rd1 end-stage retinal-degeneration mice. Stem Cell Reports 8(1):69–83. https://doi.org/10.1016/j.stemcr.2016.12.008

Chapter 16

Paired Recording to Study Electrical Coupling Between Photoreceptors in Mouse Retina

Nange Jin, Zhijing Zhang, Kimberly A. Mankiewicz, and Christophe P. Ribelayga

Abstract

Gap junction-mediated electrical coupling between retinal photoreceptors is an important determinant of photoreceptor function. Yet, quantitative measurements of the junctional conductance between coupled photoreceptors are required to fully assess the effects of coupling on visual performance. Such measurements have been obtained in salamander and other lower vertebrate retinas but are difficult to acquire in mammalian retinas, in part because of the much smaller size of photoreceptors in mammals. Here, we describe in detail a dual whole-cell patch-clamp technique we recently developed to measure the junctional conductance between photoreceptor pairs in the mouse retina. With this method, electrical coupling strength between mouse photoreceptors can be estimated with high accuracy and its impact on retinal processing of visual information further evaluated.

Key words Gap junction, Photoreceptor, Mouse, Cx36, Electrical coupling, Dual patch clamp

1 Introduction

In the central nervous system, gap junctions play an important role in cell-cell communication and signaling [1, 2]. Gap junction-mediated electrical coupling between retinal photoreceptors modulates the membrane potential of the cells and thereby many aspects of their functional properties, including their receptive field size, and the sensitivity, spectral tuning, and kinetics of their light responses [3, 4]. Work from lower vertebrates has considerably increased our knowledge of photoreceptor electrical coupling, in particular by providing measures of the junctional conductance [3, 4]. Because of the small size of mammalian photoreceptors, however, and the consequential difficulty in electrically accessing these cells, very few studies have been successful at obtaining such measures in mammals [3, 4]. Rather, many investigators, including us, have used derivative measures (such as the extent of tracer coupling, the phosphorylation state of the gap junction-forming

Chai-An Mao (ed.), *Retinal Development: Methods and Protocols*, Methods in Molecular Biology, vol. 2092, https://doi.org/10.1007/978-1-0716-0175-4_16, © Springer Science+Business Media, LLC, part of Springer Nature 2020 protein connexin, or the reliability of the rod responses) to estimate photoreceptor coupling strength [5–8]. Over the last few years, we have optimized perforated whole-cell patch-clamp recording of mouse photoreceptors [8] and have established a reliable dual technique that yields quantitative measures of the junctional conductance [9]. Here, we describe the technique in detail.

2 Materials

All solutions are prepared at room temperature using ultrapure water (Nanopure Diamond System, Thermo Fisher Scientific, Waltham, MA, USA). All procedures on animals must be in accordance with federal, local, and institutional guidelines and reviewed and approved by the local institutional animal care and use committee.

2.1 Equipment

1. A microscope (BX51WI, Olympus, Shinjuku, Japan) with a fixed stage. The microscope is equipped with two objectives: a 4× U Plan Fluorite "Universal" objective and a 40× long working distance water immersion universal M Plan Fluorite objective (Olympus, Shinjuku, Japan). The microscope is also equipped with a simple magnification changer, with 1× and 2× positions so that the maximum magnification available is $40\times \times 2\times = 80\times$.
2. For visualization of the retinal sections and pipettes, we use infrared light and differential interference contrast (DIC) and an OLY-150IR camera (Olympus) connected on the camera port of the microscope. Infrared light is provided by a 100 W halogen lamp that passes through a 900 nm long-pass filter and the condenser before illuminating the preparation from the glass bottom of the perfusion chamber.
3. A vibration cancellation table (TMC, Peabody, MA, USA) onto which the microscope is mounted.
4. A custom-made 30″ × 36″ × 40" Faraday cage mounted on the vibration cancellation table, providing full perimeter enclosure for the microscope, stage, and micromanipulators. The Faraday cage is entirely covered with black electrical field shielding plastic (Velostat, 3 M, Two Harbors, MN, USA) to prevent room lights from penetrating the cage and keeping the preparation deeply dark adapted when needed.
5. A gravity-fed perfusion system that includes a custom-made chamber of 2 mL volume, approximately 5 feet of tubing (PE-160/10, Cat# 64-0755, outer diameter, 1.57 mm; inner diameter, 1.14 mm; Warner Instruments, Hamden, CT, USA) delivering the superfusate at a flow rate of 2 mL/min or 1 turnover/min, and a temperature controller (TC-324B,

Warner Instruments, Hamden, CT, USA) that maintains the temperature in the chamber around 32 °C. The chamber is mounted on an X–Y translator for an Olympus BX51 electrophysiology microscope (Sutter Instrument, Novato, CA, USA).

6. Two patch-clamp amplifiers. We use an Axopatch 200B (Axon Instruments, Foster City, CA) and a 3900A integrating patch clamp (Dagan Corporation, Minneapolis, MN).
7. To position the microelectrodes, we use two MP-285 micromanipulators (Sutter Instrument, Novato, CA, USA) that are mounted on the fixed stage. The control units and remotes are outside the Faraday cage (*see* **Note 1**).
8. Data acquisition and recording are done using the pCLAMP Software suite and an analog/digital interface (Digidata 1440A) (Axon Instruments, Molecular Devices, Sunnyvale, CA, USA). Signals are filtered at 1 kHz with a four-pole Bessel filter and sampled at 10 kHz.
9. A micropipette puller (P-1000, Sutter Instrument) to fabricate patch pipettes.
10. Borosilicate glass capillaries (1B120F-4, outer diameter, 1.2 mm; inner diameter, 0.69; World Precision Instruments, Inc., Sarasota, FL, USA) for micropipettes.
11. A light bench that includes a 175 W xenon arc lamp, a lambda 10-3 controller unit, and a SmartShutter (Sutter Instrument). Light is forwarded to the prep through a liquid light guide (Sutter Instrument) hooked on the back port of the microscope (that is usually used for the housing of a mercury lamp) and a dichroic mirror.
12. Radiometer (ILT 1700, Peabody, MA, USA) for measuring the output of the light bench on a weekly basis and calibrating the light intensity.
13. pH Meter (pH/Ion 510, Oakton Instruments, Vernon Hills, IL, USA) for measuring and calibrating the pH values of the intracellular and extracellular solutions.
14. Osmometer (Vapro, Model 5520, Vapor Pressure Osmometer, Logan, Utah, USA) for measuring the osmolarity of the solutions.
15. A razor blade tissue chopper (Stoelting, Wood Dale, IL, USA).
16. Razor blades (Derby Extra Double Edge Razor Blades, B004SGKMA0 (544236), Istasyon Mah, Istanbul, Turkey), which need to be cleaned before use (*see* **Note 2**).
17. Liquid silicone (2021846-0807, Dow Corning Corporation, Midland, MI, USA) for attaching retinal sections to the bottom of the perfusion chamber.

2.2 Solutions

1. Intracellular (pipette) solution (in mM): 10 KCl, 120 K-D-gluconate, 5 $MgCl_2$, 5 Na_2-ATP, and 1 Na_3-GTP. The pH is adjusted to 7.25 with KOH, and osmolarity should be around 270 mosmol/L.
2. Pipette solution with 25 μM B-escin: prepare a stock solution by dissolving 1.4 mg B-escin (MP Biomedicals, Santa Ana, CA, USA) into 0.1 mL intracellular solution, and then transfer 1 μL of B-escin stock solution into 0.5 mL intracellular solution to make the final 25 μM intracellular solution for the perforated patch-clamp experiment (*see* **Note 3**).
3. Extracellular solution or perfusate: Ames' medium (Sigma-Aldrich, St Louis, MO, USA) supplemented with 23 mM $NaHCO_3$ and aerated with 5% CO_2 and 95% O_2.
4. Lucifer yellow (ammonium salt; Invitrogen, USA): dissolve 0.15 mg of Lucifer yellow into 0.2 mL intracellular solution to prepare the stock solution, and make 20 aliquots of 10 μL each. Mix a 10 μL aliquot of the Lucifer yellow stock solution into 0.3 mL intracellular solution to make the final 0.025% Lucifer yellow intracellular solution. The stock solution can be kept at −20 °C in the dark for several weeks.
5. Acetone (Fisher Scientific, Pittsburg, PA, USA) to clean the razor blades.

3 Methods

All procedures are implemented under infrared illumination. Manipulation of the animals and retinal tissue is conducted with the help of infrared goggles (D-321G-A; Night Optics USA, Huntington Beach, CA). Dissection of mouse neural retinas is performed under a dissecting microscope, under infrared illumination with dual-unit Prowler Night Vision Scopes (Meyers Electro Optics, Redmond, WA) (*see* **Note 4**).

3.1 Preparation of Mouse Retinal Slices

1. Mice are dark adapted overnight prior to being anesthetized with a mixture of ketamine and xylazine (100/10 mg/kg, intramuscular) and decapitated. One eye is enucleated and rapidly placed in bubbling perfusion solution.
2. Under the dissecting microscope, the neural retina is separated from the other ocular tissues (sclera, cornea, pigmented epithelium) by using fine tweezers. First, gently make a window in the cornea from which you will be able to open the eyeball. Second, remove the lens and pull apart the sclera. The neural retina should then isolate itself easily. Introduce a filter paper (0.45 mM HAWP; Millipore, Billerica, MA, USA) into the solution, and move the neural retina onto it. Orientate the neural retina ganglion cells down on the filter paper.

3. Move the recipient that contains the filter paper and retina bundle in solution under the chopper. Retinal slices (about 250 μm) are cut together with the filter paper and moved to and rotated 90° in the recording chamber so that the retinal layers are exposed and accessible to electrodes. Small bands of liquid silicon are used to hold the sections and the filter paper at the bottom of the chamber and in the proper orientation. The whole procedure should take less than 15 min (*see* **Note 5**).
4. The chamber with the retinal sections is then moved onto the microscope stage. Sections are perfused for 60 min to recover from surgery prior to recording (*see* **Note 6**).

3.2 Measurements of Photoreceptor Electrical Coupling

1. The two electrodes are fashioned identically (*see* **Note 7**). Electrode tips are filled with the intracellular solution and backfilled with B-escin, with or without 0.05% Lucifer yellow (Lucifer yellow is used to visualize the recorded cells; *see* Fig. 1).
2. Install the electrodes onto the micromanipulators, and move them down until they make electrical contact with the surface of the perfusion. Measure their resistance (*see* **Note 8**). The liquid junction potential is typically ~15 mV, and we do not correct for it.
3. Both electrodes are moved downward toward the retinal section at the same time, first by using the 4× objective and then by using the 40× objective and moving the magnification changer to 2× for the final approach (80× total). We use positive pressure to give the electrode tips clean access to photoreceptor membrane. Following release of the positive pressure, subsequent successful attachment to the membrane, and eventually perforation, which typically occurs within 10 min, the series resistance should be 15–30 MΩ (*see* **Note 9**).
4. To measure electrical conductance between adjacent mouse photoreceptors, we can obtain paired perforated patch-clamp recordings simultaneously from the somas of adjacent rods or the pedicles of adjacent cones or the adjacent cone pedicle and rod soma (Figs. 1 and 2). If needed, the identity of the clamped cell can be confirmed using a brief flash of light under current clamp mode (I_0).
5. To obtain measures of the junctional conductance, the voltage of one cell (slave) is held constant ($V_h = -35$ mV), and changes in membrane current are measured in response to voltage steps (50 ms, 10-mV increments from −50 to +50 mV) applied to the other cell (driver) (Fig. 2) (*see* **Notes 10–12**).

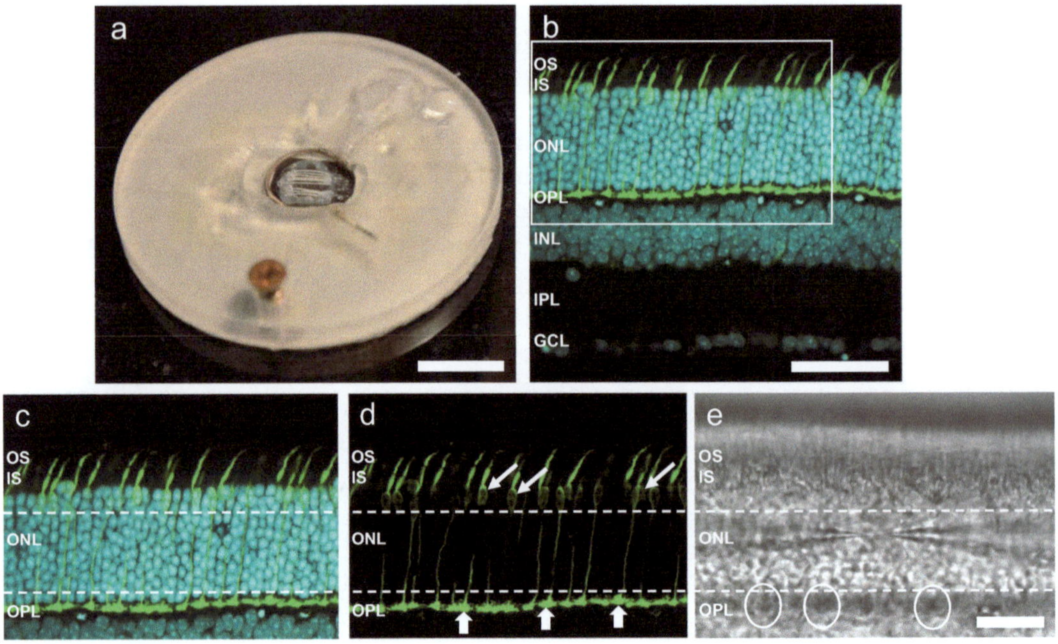

Fig. 1 Identification of photoreceptor pairs in the living mouse retinal slice. (**a**) Overview of the perfusion chamber with retinal sections in place. Note the large size of the chamber, which allows for laminar flow of the perfusion solution and ensures stability during recordings. (**b–d**) Views of a mouse retinal slice fixed in paraformaldehyde, reacted with an antibody against cone arrestin (green), and visualized under a confocal microscope. A magnified view of the outer portion (white square in [**b**]) is shown in (**c**, **d**). DAPI staining (blue) of the nuclei is shown in (**b**, **c**) and omitted in (**d**) for clarity. Note the position of the cone somas in the outermost part of the ONL (white arrows) and that of the cone pedicles at the bottom of the ONL and in the OPL (white arrowheads). (**e**) DIC image of the outer portion of a mouse retinal slice and of the recording electrodes placed on adjacent rod somas under infrared illumination in the experimental setup. Dashed lines delimit the area that includes only rod somas. This is the area targeted for recording from rods, in rod/rod or rod/cone paired recordings. To record from cones, we target the cone terminals or pedicles that are visible as contrasted structures in DIC mode (white circles). *OS* outer segments, *IS* inner segments, *ONL* outer nuclear layer, *OPL* outer plexiform layer, *INL* inner nuclear layer, *IPL* inner plexiform layer, *GCL* ganglion cell layer. Scale bars: (**a**), 2 cm, (**b**), 50 μm, (**c**)-(**e**), 25 μm. Figures (**c–e**) modified from Fig. 1 of [9], with permission

3.3 Data Analysis

A value of the current (I) is taken from the average of the values recorded between 20 and 30 ms after the beginning of the step (V). Values of the recorded current (I) are plotted as a function of the holding voltage (V). The junctional conductance is estimated from the slope of the linear regression curve that fits the experimental data (Fig. 2). Data and statistical analyses are performed with Clampfit (Molecular Devices) and OriginPro (OriginLab).

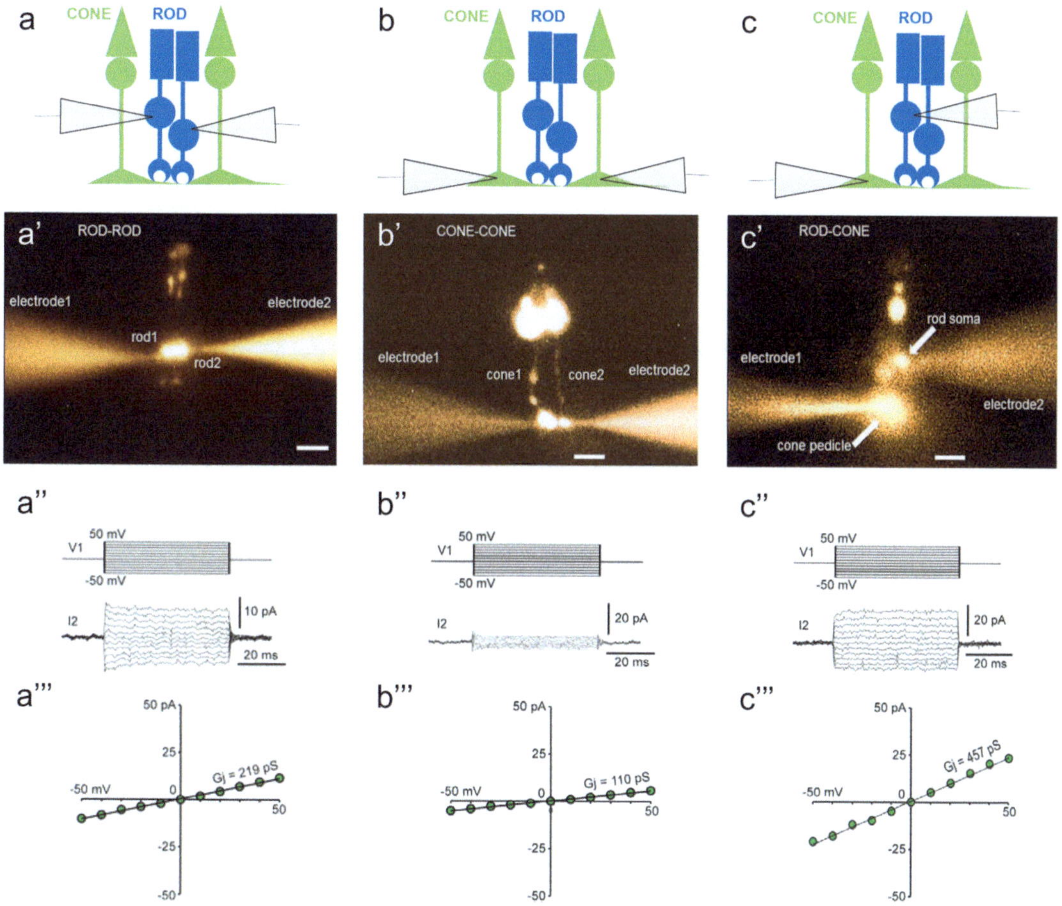

Fig. 2 Simultaneous patch-clamp recording of photoreceptor pairs in the living mouse retinal slice illustrated for a pair of rods (**a**,**a**′,**a**″,**a**‴), a pair of cones (**b**,**b**′,**b**″,**b**‴), and a rod/cone pair (**c**,**c**′,**c**″,**c**‴). (**a**–**c**) Schematic representation of the experimental setup. (**a**′–**c**′) Visualization of the photoreceptors simultaneously patch-clamped and filled with Lucifer yellow through the two recording pipettes. (**a**″–**c**″) Simultaneous voltage-clamp recording from pairs of neighboring photoreceptors. Shown are junctional currents (I_2) during a series of steps in transjunctional voltage (V_1). (**a**‴–**c**‴) Plot of $I2$ as a function of V_1. The junctional conductance (G_j) is estimated from the slope of the linear regression curve that fits the experimental data. Scale bars: (**a**′–**c**′), 5 μm. Figure (**a**′) is from Fig. 1 of [9], with authorization

4 Notes

1. Because mouse photoreceptor somata and terminals have diameters of about 1–2 μm, every possible step to reduce vibrations or movements of the preparation or equipment should be taken. These include having the micromanipulators refurbished regularly so that the electrodes do not drift during recording.
2. Razor blades need to be cleaned in acetone overnight. Rinse thoroughly with running distilled water for approximately 30 s after overnight acetone soak, and let it dry before use.

3. The pipette solution with β-escin must be prepared fresh (directly prior to the start of an experiment). Also, β-escin powder needs to be stored under dry conditions and in the dark (it is light sensitive) in order to prevent its transformation into α-escin, which is the inactive form.
4. Current lighting conditions or immediate previous light history, as well as circadian time and mouse strain, may affect the gap junctional conductance between photoreceptors [5–9]. It is thus extremely important to control the lighting conditions during an experiment. Typically, we keep the animals in the dark prior to an experiment and perform the surgery under deep dark-adapted conditions (infrared light). If light adaptation is required, we apply a background of light from the light bench, under the microscope.
5. Sections should never be moved outside of the solution and exposed to air. We use a container in which to isolate and section the neural retina that is large enough to accommodate the chamber so that sections can be moved to the chamber without being moved out of the solution.
6. When the chamber is placed under the microscope and during the 60 min recovery period, use that time to search for the perfect area for the recordings. In particular, when recording from cone pedicles, search for an area where these structures are clearly visible. Spending time arranging the slices and finding areas where cells and terminals are clearly visible will increase the success rate of your recordings.
7. The ideal patch pipette geometry is a rapid taper to the tip so that the shank length is short, thereby minimizing resistance to the pipette tip. Settings used to achieve this geometry vary with puller type, platinum filament characteristics, and glass diameter. Generally, 4–5 steps of pulling are required to achieve a high degree of tapering at the shoulder of the pipette.
8. Pipette resistance should be around 15 MΩ [10]. Discard the pipettes if the resistance is smaller than 12 MΩ or larger than 18 MΩ.
9. The seal resistance should be >1 GΩ. If the seal resistance is less than 1 GΩ or the series resistance (sum of the pipette resistance and resistance imposed by low conductivity elements such as membrane fragments that are sucked in the pipette during the whole cell formation process) is larger than 30 MΩ, discard the cell.
10. This protocol minimizes the influence of voltage-gated channels on the measurements. In particular, holding the driver cell at 0 mV minimizes the dark current and activation of the voltage-gated potassium and calcium channels. In addition, we restrict the voltage steps to between −50 and +50 mV to

minimize activation of the hyperpolarization-activated current (I_h), which opens at potentials negative to −60 mV. Finally, using brief voltage steps minimizes the contribution of calcium-activated chloride channels that develop with a time constant of activation ≫50 ms. We typically check the linearity of the current-voltage relationship for each recorded pair. The near-perfect linearity of the current-voltage relationship we obtain (*see* Fig. 2) indicates that the contribution of the voltage-gated conductances to our measurements is indeed minimal. If contamination from voltage-dependent conductances or if any current leaks are observed, the data are discarded.

11. Troubleshooting: if the photovoltage is unusually small, cell attachment or giga seal is difficult to establish, or if you cannot hold a cell for more than 1 min, check the following:
    (a) Check if the vibration table is stable (also, *see* **Note 1**).
    (b) Check if bubbling of the perfusion is still occurring. Stopping bubbling with 5% CO_2/95% O_2 will have rapid consequences on the pH of the solution and survival of the retinal cells.
12. Dealing with the noise of the system: junctional currents between mammalian photoreceptors are typically small (i.e., few hundreds of ps). To obtain accurate measures of these currents, we need to evaluate and subtract the background noise of the system. The background noise can be measured when the two electrodes are in the bath and not touching any cells. This current is also independent of the position of the recording electrodes (e.g., identical whether the electrodes were 2, 5, or 10 cells apart) and insensitive to the gap junction antagonist meclofenamic acid [8, 11]. The amplitude of the noise in the system is typically between 50 and 100 ps (see, for instance, [12, 13]) and has to be subtracted from the recordings. Lastly, another type of noise is a high-frequency fluctuation of the baseline of recorded currents. If this occurs, re-chlorination of the silver wire of the Ag/AgCl bridge may be required.

Acknowledgments

This work was supported by NIH grants EY018640, EY029408, and EY028102 and the Hermann Eye Fund.

References

1. Connors BW (2017) Synchrony and so much more: diverse roles for electrical synapses in neural circuits. Dev Neurobiol 77:610–624
2. Nagy JI, Pereda AE, Rash JE (2018) Electrical synapses in mammalian CNS: past eras, present focus and future directions. Biochim Biophys Acta Biomembr 1860:102–123
3. Bloomfield SA, Volgyi B (2009) The diverse functional roles and regulation of neuronal gap junctions in the retina. Nat Rev Neurosci 10:495–506
4. Ribelayga CP, O'Brien J (2018) Circadian and light-adaptive control of electrical synaptic plasticity in the vertebrate retina. In: Jing J (ed) Network functions and plasticity: perspectives from studying electrical coupling in microcircuits. Elsevier, Cambridge. in press
5. Ribelayga C, Cao Y, Mangel SC (2008) The circadian clock in the retina controls rod– cone coupling. Neuron 59:790–801
6. Li H, Zhang Z, Blackburn MR, Wang SW, Ribelayga CP, O'Brien J (2013) Adenosine and dopamine receptors coregulate photoreceptor coupling via gap junction phosphorylation in mouse retina. J Neurosci 33:3135–3150
7. Jin NG, Chuang AZ, Masson PJ, Ribelayga CP (2015) Rod electrical coupling is controlled by a circadian clock and dopamine in mouse retina. J Physiol 593:1597–1631
8. Zhang Z, Li H, Liu X, O'Brien J, Ribelayga CP (2015) Circadian clock control of connexin36 phosphorylation in retinal photoreceptors of the CBA/CaJ mouse strain. Vis Neurosci 32: E009
9. Jin NG, Ribelayga CP (2016) Direct evidence for daily plasticity of electrical coupling between rod photoreceptors in the mammalian retina. J Neurosci 36:178–184
10. Jin NG, Chuang AZ, Masson PJ, Ribelayga CP (2015) Reply from Nan Ge Jin, Alice Z. Chuang, Philippe J. Masson and Christophe P. Ribelayga. J Physiol 593:2977–2978
11. Pan F, Mills SL, Massey SC (2007) Screening of gap junction antagonists on dye coupling in the rabbit retina. Vis Neurosci 24:609–618
12. Li W, DeVries SH (2004) Separate blue and green cone networks in the mammalian retina. Nature Neurosci 7:751–756
13. Li PH, Verweij J, Long JH, Schnapf JL (2012) Gap-junctional coupling of mammalian rod photoreceptors and its effect on visual detection. J Neurosci 32:3552–3562

Index

A

B

C

Chai-An Mao (ed.), *Retinal Development: Methods and Protocols*, Methods in Molecular Biology, vol. 2092, https://doi.org/10.1007/978-1-0716-0175-4, © Springer Science+Business Media, LLC, part of Springer Nature 2020

H

I

J

K

L

M

N

O

MIX
Papier aus verantwortungsvollen Quellen
Paper from responsible sources
FSC® C105338

If you have any concerns about our products,
you can contact us on
ProductSafety@springernature.com

In case Publisher is established outside the EU,
the EU authorized representative is:
Springer Nature Customer Service Center GmbH
Europaplatz 3, 69115 Heidelberg, Germany

Printed by Libri Plureos GmbH
in Hamburg, Germany